METHODS IN MOLECULAR BIOLOGY

Series Editor
John M. Walker
School of Life and Medical Sciences
University of Hertfordshire
Hatfield, Hertfordshire, UK

For further volumes:
http://www.springer.com/series/7651

For over 35 years, biological scientists have come to rely on the research protocols and methodologies in the critically acclaimed *Methods in Molecular Biology* series. The series was the first to introduce the step-by-step protocols approach that has become the standard in all biomedical protocol publishing. Each protocol is provided in readily-reproducible step-by-step fashion, opening with an introductory overview, a list of the materials and reagents needed to complete the experiment, and followed by a detailed procedure that is supported with a helpful notes section offering tips and tricks of the trade as well as troubleshooting advice. These hallmark features were introduced by series editor Dr. John Walker and constitute the key ingredient in each and every volume of the *Methods in Molecular Biology* series. Tested and trusted, comprehensive and reliable, all protocols from the series are indexed in PubMed.

Vascular Morphogenesis

Methods and Protocols

Second Edition

Edited by

Domenico Ribatti

Department of Basic Medical Sciences, Neurosciences and Sensory Organs, University of Bari Medical School, Bari, Italy

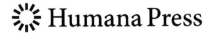 Humana Press

Editor
Domenico Ribatti
Department of Basic Medical Sciences
Neurosciences and Sensory Organs
University of Bari Medical School
Bari, Italy

ISSN 1064-3745   ISSN 1940-6029   (electronic)
Methods in Molecular Biology
ISBN 978-1-0716-0915-6   ISBN 978-1-0716-0916-3   (eBook)
https://doi.org/10.1007/978-1-0716-0916-3

This Humana imprint is published by the registered company Springer Science+Business Media, LLC, part of Springer
Nature.
The registered company address is: 1 New York Plaza, New York, NY 10004, U.S.A.

Preface

Development of the vascular system involves a complex sequence of inductive and differentiating signals leading to vasculogenesis and/or angiogenesis. Vascular networks formed by vessel sprouting undergo extensive vascular remodeling to form a functional, organotypically differentiated vasculature. Blood vessels form one of the body's largest surfaces, serving as a critical interface between the circulation and the different organ environments. Dissecting and exploring this process in its multifaceted morphological and molecular aspects has represented a basic contribution and a fascinating adventure in the history of biology.

Several genetic and epigenetic mechanisms are involved in the early development of the vascular system, and there is an extensive literature on the genetic background and the molecular mechanisms responsible for blood vessel formation. Evidence is now emerging that blood vessels themselves have the ability to provide instructive regulatory signals to surrounding nonvascular target cells during organ development. The recently established concept of angiocrine signaling can explain how endothelial cells shape their microenvironment. While microvascular endothelial cells were originally believed to form an inert cellular population, they are now believed to actively control the tissue microenvironment by endothelial cell-derived angiocrine factors, which control organogenesis as well as organ function in health and disease. Thus, endothelial cell signaling is currently understood to promote fundamental cues for cell fate specification, embryo patterning, organ differentiation, and postnatal tissue remodeling. The microvasculature displays remarkable structural and functional diversity. Understanding the concept of vascular bed specificity represents a major challenge for future investigations. Indeed, one of the most interesting theoretical perspectives and practical applications of endothelial cell signaling is the possibility for these cells to maintain their inductive potential during adult life.

The aim of the second edition of this book is to provide a range of methods and protocols for studying vascular morphogenesis in vivo and in vitro to reflect advances in the field. I hope that this book attracts a wide audience among cell biologists, anatomists, pathologists, and physiologists, and that the reader finds this book instructive and useful.

I would also like to thank the colleagues who kindly agreed to contribute to this book.

Bari, Italy   *Domenico Ribatti*

Contents

Contributors

RIEKO ASAI • *Cardiovascular Research Institute, University of California San Francisco, San Francisco, CA, USA; Department of Physiological Chemistry and Metabolism, Graduate School of Medicine, University of Tokyo, Tokyo, Japan*

ELISA BOSCOLO • *Division of Experimental Hematology and Cancer Biology, Cincinnati Children's Hospital Medical Center, Cincinnati, OH, USA; Department of Pediatrics, University of Cincinnati College of Medicine, Cincinnati, OH, USA*

MICHAEL BRESSAN • *Department of Cell Biology and Physiology, McAllister Heart Institute, University of North Carolina at Chapel Hill, Chapel Hill, NC, USA*

ANNELIES BRONCKAERS • *Faculty of Medicine and Life Sciences, Biomedical Research Institute (BIOMED), UHasselt—Hasselt University, Diepenbeek, Belgium*

VALERIO CICCONE • *Department of Life Sciences, University of Siena, Siena, Italy*

ANCA MARIA CIMPEAN • *Department of Microscopic Morphology/Histology, Angiogenesis Research Center Timisoara, Victor Babes University of Medicine and Pharmacy, Timisoara, Romania*

MARIELLA ERREDE • *Human Anatomy and Histology Unit, Department of Basic Medical Sciences, Neurosciences and Sensory Organs, Bari University School of Medicine, Bari, Italy*

ARIANNA GIACOMINI • *Unit of Experimental Oncology and Immunology, Department of Molecular and Translational Medicine, University of Brescia, Brescia, Italy*

FRANCESCO GIROLAMO • *Human Anatomy and Histology Unit, Department of Basic Medical Sciences, Neurosciences and Sensory Organs, Bari University School of Medicine, Bari, Italy*

ELLEN GO • *Division of Pediatric Rheumatology, The Hospital for Sick Children, Toronto, ON, Canada*

JILLIAN GOINES • *Division of Experimental Hematology and Cancer Biology, Cincinnati Children's Hospital Medical Center, Cincinnati, OH, USA*

DIEGO GUIDOLIN • *Section of Anatomy, Department of Neuroscience, University of Padova Medical School, Padova, Italy*

DAVID GUREVICH • *School of Biochemistry, University of Bristol, Bristol, UK*

PETRA HILKENS • *Faculty of Medicine and Life Sciences, Biomedical Research Institute (BIOMED), UHasselt—Hasselt University, Diepenbeek, Belgium*

JANOS M. KANCZLER • *Bone and Joint Research Group, Centre for Human Development, Stem Cells and Regeneration, Faculty of Medicine, University of Southampton, Southampton, UK*

HWAN D. KIM • *Department of Cardiac Surgery, Boston Children's Hospital, Boston, MA, USA; Department of Surgery, Harvard Medical School, Boston, MA, USA*

IVO LAMBRICHTS • *Faculty of Medicine and Life Sciences, Biomedical Research Institute (BIOMED), UHasselt—Hasselt University, Diepenbeek, Belgium*

IRINA V. LARINA • *Department of Molecular Physiology and Biophysics, Baylor College of Medicine, Houston, TX, USA*

RUEI-ZENG LIN • *Department of Cardiac Surgery, Boston Children's Hospital, Boston, MA, USA; Department of Surgery, Harvard Medical School, Boston, MA, USA*

ANDREW L. LOPEZ III • *Department of Molecular Physiology and Biophysics, Baylor College of Medicine, Houston, TX, USA*

JUAN M. MELERO-MARTIN • *Department of Cardiac Surgery, Boston Children's Hospital, Boston, MA, USA; Department of Surgery, Harvard Medical School, Boston, MA, USA; Harvard Stem Cell Institute, Cambridge, MA, USA*

HARRY MELLOR • *School of Biochemistry, University of Bristol, Bristol, UK*

TAKASHI MIKAWA • *Cardiovascular Research Institute, University of California San Francisco, San Francisco, CA, USA*

LUCIA MORBIDELLI • *Department of Life Sciences, University of Siena, Siena, Italy*

RICHARD O. C. OREFFO • *Bone and Joint Research Group, Centre for Human Development, Stem Cells and Regeneration, Faculty of Medicine, University of Southampton, Southampton, UK*

MARCO PRESTA • *Unit of Experimental Oncology and Immunology, Department of Molecular and Translational Medicine, University of Brescia, Brescia, Italy*

MARIUS RAICA • *Department of Microscopic Morphology/Histology, Angiogenesis Research Center Timisoara, Victor Babes University of Medicine and Pharmacy, Timisoara, Romania*

DOMENICO RIBATTI • *Department of Basic Medical Sciences, Neurosciences and Sensory Organs, University of Bari Medical School, Bari, Italy*

MARCO RIGHI • *Consiglio Nazionale delle Ricerche, Institute of Neuroscience, Milano, Italy; NeuroMI Milan Center for Neuroscience, University of Milano-Bicocca, Milan, Italy*

AMBER N. STRATMAN • *Division of Developmental Biology, Eunice Kennedy Shriver National Institute of Child Health and Human Development, National Institutes of Health, Bethesda, MD, USA; Department of Cell Biology and Physiology, Washington University School of Medicine, St. Louis, MO, USA*

RYO SUDO • *School of Integrated Design Engineering, Graduate School of Science and Technology, Keio University, Yokohama, Japan; Department of System Design Engineering, Faculty of Science and Technology, Keio University, Yokohama, Japan*

ANANTHALAKSHMY SUNDARARAMAN • *School of Biochemistry, University of Bristol, Bristol, UK*

JACOB THASTRUP • *2cureX A/S, Copenhagen, Denmark*

CINZIA TORTORELLA • *Section of Anatomy, Department of Neuroscience, University of Padova Medical School, Padova, Italy*

SARAH LINE BRING TRUELSEN • *2cureX A/S, Copenhagen, Denmark*

DANIELA VIRGINTINO • *Human Anatomy and Histology Unit, Department of Basic Medical Sciences, Neurosciences and Sensory Organs, Bari University School of Medicine, Bari, Italy*

MASAFUMI WATANABE • *School of Integrated Design Engineering, Graduate School of Science and Technology, Keio University, Yokohama, Japan*

HAOCHE WEI • *School of Biochemistry, University of Bristol, Bristol, UK*

BRANT M. WEINSTEIN • *Division of Developmental Biology, Eunice Kennedy Shriver National Institute of Child Health and Human Development, National Institutes of Health, Bethesda, MD, USA*

JULIA A. WELLS • *Bone and Joint Research Group, Centre for Human Development, Stem Cells and Regeneration, Faculty of Medicine, University of Southampton, Southampton, UK*

ESTHER WOLFS • *Faculty of Medicine and Life Sciences, Biomedical Research Institute (BIOMED), UHasselt—Hasselt University, Diepenbeek, Belgium*

MERVIN C. YODER • *Indiana Center for Regenerative Medicine and Engineering, Indiana University School of Medicine, Indianapolis, IN, USA*

MARINA ZICHE • *Department of Medicine, Surgery and Neurosciences, University of Siena, Siena, Italy*

## Chapter 1

Historical Overview of In Vivo and In Vitro Angiogenesis Assays

Anca Maria Cimpean and Marius Raica

Abstract

Most of angiogenesis assays were designed and developed during Folkman's era. But the growth of new blood vessels in several pathologic conditions as tumor development or inflammation were observed long time ago.

The development of new blood vessels was early observed by ancient Egyptians who tried to destroy them by applying empirical methods. From the first observations regarding angiogenesis to a personalized therapy targeting newly formed blood vessels a lot of experimental in vitro and in vivo angiogenesis assays have been developed. The present work will overview the oldest and less known part of angiogenesis assays development, and in addition, it will present the newest data in the experimental field of angiogenesis which is rapidly improved by the needs of new antiangiogenic and antivascular therapy development.

Key words History, Angiogenesis, Assays

1 The Past Will Always Govern the Present!

The earliest evidences of angiogenesis are found in Edwin Smith Papyrus (seventeenth century BCE) followed by Ebers Papyrus (sixteenth century BCE) which include data regarding the diagnosis of the tumors together with a description of "vessel-tumor." Tumors were considered to arise from the wound of the vessel. Also, in the same papyrus, there is a mention of what we could consider as the first antiangiogenic therapy, somewhat a surgical one, carried out by the use of a knife that has been heated in a fire in order to destroy the vessels surrounding the tumor and to reduce the amount of bleeding [1]. Thus, ancient Egyptians had already observed the high vascularization of the tumor tissue and tried to treat it using empirical methods.

Later, Hippocrates (460–370 BC) observed the blood vessels around a malignant tumor and described them as a structure resembling the claws of a crab. He named the disease *karkinos* (the Greek name for crab) based on the arrangement of blood

Domenico Ribatti (ed.), *Vascular Morphogenesis: Methods and Protocols*, Methods in Molecular Biology, vol. 2206, https://doi.org/10.1007/978-1-0716-0916-3_1, © Springer Science+Business Media, LLC, part of Springer Nature 2021

vessels around the tumor and not because of the morphology of the tumor. In a modern era of angiogenesis, these radial arrangement of blood vessels like a spoke wheel was described in the chick embryo chorioallantoic membrane model around tumor implants.

Aristotle was the first scholar who performed extensive studies on living world as a separate field of inquiry and moreover he studied a 40-day-old human fetus. His experiments suggested that heart along with blood vessels was the first structure to appear during embryogenesis [2]. By this early description, Aristotle may be considered the first scientist who mentioned vasculogenesis and early steps of normal angiogenesis during embryonic and fetal stage of human body development.

Herophilus of Chalcedon (300 BC) performed human cadaver dissection, elaborated distinction between arteries and veins, and discovered that arteries contained blood rather than air. He is considered the founder of scientific methods and experimental assays on humans. Herophilus's work was continued thereafter by experiments of Michael Servetus, Ibn al-Nafis, Amato Lusitano, or William Harvey.

The first use of term "capillary" for the most minute of blood vessels is attributed to the architect, engineer, scientist, inventor, poet, sculptor, painter and anatomic artist Leonardo da Vinci. He became interested in anatomy due to his collaboration with Veronese anatomist named Marc Antonia Della Torre. Because of the unexpected death of Della Torre, da Vinci took over also the dissection techniques. Da Vinci injected the blood vessels with wax for preservation, an anatomical technique still used today, and in so doing he discovered and named the capillaries (although he did not understand the role they played connecting the arterioles and venules).

William Harvey published in 1628 *Exercitatio anatomica de motu cordis et sanguinis,* where he suggested the presence of capillaries. For this purpose, he applied empirical experimental methods as the boil of organs, including the liver, spleen, lungs, and kidney, and then dissected the tissue until he could see "capillamenta," or capillary threads [3]. His idea of the existence of the capillary bed would be confirmed 30 years later by scientists with the aid of microscopy.

In 1651, Harvey made early experiments by using chick embryos to prove the origin of the body from an egg. On this animal model he described for the first time the yolk sac and he noticed that "islands" of blood cells formed before the heart did [4].

The microscope represented a big step and challenge for research and the vascular system was deeply studied by using such an instrument. Several visualization techniques appeared and they needed to be improved by additional methods as the use of special stainings and dyes.

Jean Riolan (1580–1657) injected for the first time color dyes to demonstrate the branching of the vascular tree [5]. Other dyes as saffron, carmine, Prussian blue, India ink, and silver nitrate in aqueous or gelatin suspension [6] have been used in anatomy and microscopy. Starch, plaster of Paris, glue, asphalt, latex or rubber, gum Arabic, sodium silicate, oil of sesame, shellac, thymol, and mercury were injected into the vascular system [6]. Nowadays, latex, for example, is still widely used in angiogenesis experimental models ([7–9]. Jan Swammerdam injected color wax into arteries and veins. He also discovered and described lymphatic vessel valves.

Marcello Malpighi contributed a lot to the development of angiogenesis assays by describing capillaries in the mesentery and the lung of the frog. By the use of microscopic techniques he studied the vasculature of tadpoles, fish, rooster combs, rabbit ears, and bat wings and he accurately described the vasculature of chick embryo chorioallantoic membrane—a widely used experimental model in angiogenesis field—in his work *De ovo incubato* in 1671. Anatomic injection of vascular tree was extensively performed by Malpighi who used ink, urine colored with ink, and black-colored liquid mixed with wine. He described by this method the capillaries of the renal glomeruli as "a beautiful tree loaded with apples." Frederik Ruysch (1638–1731) improved vascular injection by use of wax and proved the presence of capillary bed in all tissues including vasa vasorum and bronchial capillaries [3].

The main cellular player of angiogenesis, the endothelium, was attributed to be mentioned for the first time by Theodor von Schwann, the proponent of the cell theory, who characterized it as follows: "very distinct cell-nuclei occur at different spots upon the walls of the capillaries ... they are either the nuclei of the primary cells of the capillaries, or nuclei of epithelial cells, which invest the capillary vessels ... these nuclei frequently seemed to lie free upon the internal wall of the vessel ... that these are the nuclei of the primary cells of the capillaries is, therefore, most probable" [10]. These data were sustained by application of silver nitrate injection into the vessels followed by the light exposure which was performed by several German investigators [11]. This method definitively established the presence of a cellular lining and hence the existence of a capillary wall. After the vessels were injected with the silver solution and the tissue was exposed to light, a dark brown precipitate lining the cell boundaries was observed. In 1862, von Recklinghausen stained the lymphatics with this method and was the first who noted that they were lined with cells [12]. Later, a German group added gelatin to silver nitrate in order to maintain blood vessels open.

Swiss anatomist Wilhelm His named for the first time the inner layer of the blood vessels as endothelium in his work entitled *Die Häute und Höhlen des Körpers* (The Membranes and Cavities of the Body) [13, 14].

But the most controversial cells of the capillary wall are mural cells, which have an interesting history. Despite the fact that Lister suggested the contractility of the capillaries in 1858 and Roy and Brown used an interesting apparatus [15] that looked similar to the future transparent chamber of Sandison, successful visualization of mural cells had not been made possible. In 1873, Rouget described a kind of cells with branched morphology and irregular cell bodies having contractile properties that encircled capillaries.

Scientific world at the time totally disagreed with Rouget's papers about the mural cells and soon they were completely forgotten [16]. Rouget did not use a stain to highlight mural cells, but Zimmerman in 1886 (with silver nitrate) and Mayer in 1902 (methylene blue) stained these perivascular cells [17]. Also, Zimmerman used the name "pericyte" for Rouget cells and postulated that contraction of these cells controls capillary permeability [18].

In his book "Anatomy and Physiology of the capillaries," Krough [15] described his experiments on the effects of pituitary extracts on capillaries which probably previewed the future assays of Gospodarowicz (1975, 1989), isolation and purification of bovine pituitary fibroblast growth factor [19], Ferrara's crucial VEGF discovery and isolation of microvascular endothelial cells for in vitro assays by Folkman [20].

The embryological development of blood vessels was first mentioned by Wilhelm Roux in his doctoral thesis [21]. The bifurcation of blood vessels described by Roux represented the first evidence to differentiate between angiogenesis and arteriogenesis [22]. Roux had also made a big contribution to the field of tissue culture by isolating medullary plate of the chicken embryo in 1885 and maintaining it in a warm saline solution for several days. This medium was later used for ring aortic angiogenesis assays and many other studies with the same principle [23, 24].

A further step in the history of angiogenic assays is attributed to people who observed that pathologic tissues like tumors or inflammatory specimens are macroscopically rich in blood vessels.

John Hunter recognized the development of new blood vessels in healing wounds [25]. In the late 1800s, Virchow [26] and other German pathologists observed that some human tumors are highly vascularized. These findings were confirmed later by Goldman [27] and Warren Lewis [28] who described many differences in the morphology of blood vessels among various tumors in rats and humans. With a tumor window model Gordon Ide and colleagues [29] observed a potent angiogenic response after implanting a tumor in a rabbit's ear and thus postulate the presence of tumor derived vascular growth factors which they called "vessel growth-stimulating substance." Clark, in his lab, used the same experimental model [30, 31] to study the development and function of blood vessels and also lymphatics [32]. Glenn Algire correlated the number of blood vessels with tumor size and proposed that tumor neovascularization represents an essential step in tumorigenesis. He and his team explained a detailed mechanism of the vascular response to tumor transplants as follows: "the growth advantage of a tumour cell over its normal counterpart might not be owing to some hypothetical capacity for autonomous growth inherent within the tumour cell, but rather to its ability to continuously induce angiogenesis" [33]. Greenblatt and Shubik interposed a Millipore filter between the tumor and the host stroma and observed that the angiogenic substance previously described by Ide is a diffusible factor which can be theoretically identified [34]. Three years later, Judah Folkman proved its existence as TAF, proposing that tumor growth might be prevented if TAF activity is blocked [35].

Weibel–Palade bodies store von Willebrand factor inside of what Ewald Weibel and George Emil Palade described as "unknown rod-shaped cytoplasmic component which consists of a bundle of fine tubules, enveloped by a tightly fitted membrane, regularly found in endothelial cells of small arteries in various organs in rat and man". The abstract of their paper ended thus: "The nature and significance of these cytoplasmic components are yet unknown". Today it is well-known that one component of these bodies is von Willebrand factor, stored in these specific organelles of endothelial cells. FVIII-related antigen antibodies are used today for identification of endothelial cell differentiation in vitro and immunostaining of blood vessels in normal and tumor tissues samples [36].

Many data about angiogenesis derived from pioneering studies of neovascularization of regenerating tissues. For example, the sprouting model of angiogenesis—described by Ausprunk in 1979—originated from studies made in different transparent essentially two-dimensional experimental wounds, from those of frog's webs or bat wings performed by Wharton-Jones [37] to the rabbit ear chamber [31]. Early studies of angiogenesis were performed by direct observations on live specimens. Using a transparent chamber introduced into a tissue defect, Clark described, in exhaustive detail, the growth of a solid cord of endothelium, from the convex side of a curved vessel that continued to grow and anastomose with other endothelial cords or with preexisting vessels. Since mitotic figures were observed at the proximal end of the sprout, it was concluded that the sprout advanced by the migration of dividing endothelial cells at the proximal end of the tip [38], an observation that would be confirmed over 40 years later by Ausprunk and Sholley [39, 40].

2 Modern Era of Angiogenesis Assays

The last book "Angiogenesis An Integrative Approach From Science To Medicine," published by Folkman and Figg before Judah Folkman's death stated in the first chapter that "the lack of bioassays for angiogenesis, the inability to culture endothelial cells in vitro in the early 1970s, and the absence of any angiogenesis regulatory molecules did not help matters [41]. Before angiogenic or antiangiogenic molecules could be found, it was necessary to develop bioassays for angiogenesis."

Probably, Folkman and his team had the highest experience into development of several angiogenesis assays. Indeed, the number of models for angiogenesis grew exponentially and, recently a group of researchers in angiogenesis published consensus guidelines for using angiogenesis assays [42]. The consensus guidelines paper may be considered in this moment as a crucial work in the field of angiogenesis assays.

Morphogenesis of the endothelium. Werner Risau was a pioneer in the study of basic vascular morphogenesis and its induction in several experimental studies. Accurate and extensive studies concerning mechanism of both vasculogenesis [43] and angiogenesis [44] have been developed and embryonic angiogenic factors identified [45]. Brain–blood barrier structure represented one of the favorite issues of Werner Risau. Together with his collaborators, he characterized a monoclonal antibody named HT7 which is expressed specifically on cerebral endothelium in chick. They described the induction of HT7 antigen in endothelial cells of chick chorioallantoic vessels (which normally do not express this protein) after embryonic mouse brain tissue implantation [46–48].

Tumor growth in isolated perfused organ model. Judah Folkman is considered the father of angiogenesis and he had an accurate, fine clinical sense [49]. Ten years before 1971 when angiogenesis-dependent tumor growth concept was postulated, Judah Folkman and Frederick Becker performed perfusion of hemoglobin solutions into the carotid artery of rabbit and canine thyroid glands isolated in glass chambers while studying hemoglobin as a possible blood substitute [50, 51]. They observed that implanted mouse melanoma cells developed as a tumor which grew until a size of no more than 1 mm^3. Implanted on syngeneic mice the same tumors expanded and grew more than 1000 times compared with their original volume in the perfused thyroid gland. This experiment suggested that tumors are angiogenesis dependent.

2.1 Corneal Neovascularization Assay

In 1969, the angiogenesis-dependent tumor growth was sustained by Folkman clinical observation of one retinoblastoma from a child eye. The tumor which protruded from retina to vitreous had a big size and was highly vascularized. Small numerous tumors were observed in the anterior chamber of the eye; all of them were no more than 1.25 mm. This was the first evidence which supported a starting point for developing of a new model to study tumor angiogenesis—corneal neovascularization model [41]. In the early 1970s, Folkman and Gimbrone implanted tumor fragments with a size of no more than 0.5 mm^3 into rabbit cornea at a distance of

2 mm from the limbal edge. After 8–10 days new capillary blood vessels grew from the limbus, invaded the cornea, and encircled the tumor implant in the absence of inflammation and vascularized tumor grew exponentially [52, 53]. This experiment demonstrated the presence and action of a tumor angiogenic diffusible factor which stimulated the recruitment of new blood vessels by the tumor. When the experiment was reedited by replacement of the tumor with a tumor extract the high diffusibility of the tumor extract failed to produce an angiogenic response on corneal angiogenesis model. This impediment was removed by Robert Langer which combined a polymer with lyophilized protein to be tested [54, 55]. The rabbit model of corneal angiogenesis assay was later adapted for mouse corneal micropocket [56]. Molecular regulation of angiogenesis, development of new inhibitory drugs for angiogenesis, and regression of newly formed blood vessels after removal of the angiogenic stimulus represent few results of the use of this model in angiogenesis research.

Endothelial cell culture. In vitro studies. The first successful growth and passage of vascular endothelial cells in vitro (from human umbilical veins) by Gimbrone et al. [52, 53] in Folkman's lab [52, 53] and Jaffe in his laboratory from Cornell [57] represented the beginning of the development of long-term in vitro cultures of vascular endothelial cells. The first long-term passage of cloned capillary endothelial cells came later [58]. Donald Ingber made an important contribution in this field by demonstration of the fact that cultured endothelial cells changes in cell shape can signal through integrins to regulate gene expression and DNA synthesis. He went on to develop an entirely new field of investigation of cell biology based on the mechanisms by which mechanical forces modify DNA synthesis and gene expression [59].

Angiogenesis was first observed in vitro by Folkman and Haudenschild [60]. After long-term culture of capillary endothelial cells, these authors observed the spontaneous organization of these cells into capillary-like structures. The development of lumen from vesicles formed inside the cultured endothelial cells and of tubes and branches from endothelial monolayers are observed in this system. Bidimensional and 3D variants of endothelial cell cultures tried to mimic the in vivo development of blood vessels. Roberto Montesano has contributed to clarifying some cellular and molecular mechanisms of angiogenesis and tubulogenesis using an original three-dimensional cell culture system which replicates key events of angiogenesis and tubulogenesis, thereby facilitating molecular analysis [46–48].

In Alexis Carrel and Alexander Maximow studies, endothelial cells survived in explants that retained three dimensional organotypic organizations [61]. Early observations made by Maximow [62] on the growth of what was claimed to be endothelium in tissue culture described an initial outgrowth of endothelium-like cells from vascular explants, followed by an outgrowth of spindle-shaped cells. At this stage the endothelium-like appearances were lost, although claims were made for the appearance of structures reminiscent of sprouting capillaries [62].

Aortic ring explants developed later by Nicosia and its variants is a valuable modern extension of these classic studies [63]. From this study originated cell migration assays in the three-dimensional organ cultures. After the development of the stable and long-term endothelial cell culture it followed the use chronologically the macrovascular endothelial cells, from human umbilical vein or bovine aorta, microvascular endothelial cells use and also the use of numerous proliferation, migration, tube formation assays or protease assays.

2.2 In Vitro Vasculogenesis Assays

The initial steps of vasculogenesis were studied by Flamme in the early 1990s [64]. Embryo-derived mesodermal cell culture and embryonic stem (ES) cell differentiation assays enable researchers to investigate vasculogenesis virtually as it occurs in the embryo in vivo. Adherent cultures of dissociated cells from quail blastodiscs have been reported to generate both hematopoietic and endothelial cells that aggregate into characteristic blood islands and give rise to vascular structures in long-term culture. This process proved to be strictly dependent by FGF-2 [65]. The same quail blastodisc cells were grown in suspension and formed three-dimensional spherules which developed vasculogenic processes, as described by Krah et al. in 1994 [66]. In the past decades, the most promising systems is the murine ES cell–derived embryoid body formation assays. This model system, whereby a primitive vascular plexus is formed, provides an attractive tool for dissecting the mechanisms involved in the vasculogenesis process: angioblast differentiation, proliferation, migration, endothelial cell–cell adhesion, and vascular morphogenesis can all be evaluated.

Chick embryo chorioallantoic membrane assay is one of the earliest models used to grow xenografts. Investigations on the distributions and branching of blood vessels of the vascular membrane of chick embryo at various ages were made by Popoff [67] and Virchow. The early stages of avian vascular system development were described by Sabin in 1920 [68].

In 1911, Rous and Murphy demonstrated the growth of the Rous 45 chicken sarcoma transplanted onto the CAM [69]. Later, Murphy reported that mouse and rat tumors implanted onto the CAM could be maintained by continuous passage from egg to egg and described the effects of this heterologous transplantation on CAM and chick embryo [70]. Murphy tried to culture and passage human tumors but was less successful [46–48]. Based on these findings, Bischop and Varmus [71] went on to study the gene product of *src* in mammals that was nearly identical to a sequence in the normal cellular DNA of several different species of birds.

Whole or partial vessel outgrowth assay. Initially developed by Nicosia and Ottinetti in 1990 [63] the rat ring aortic angiogenesis assay was subsequently adapted for other species. It followed the development and adaptation of this test for mouse, chicken or bovine ring aortic assays with observation of the differences between the behaviors of the endothelial cells from different species in various culture media [72]. The need to establish human models of angiogenesis resulted in the development of a humanized form of ring assays using tissues from placental blood vessels and saphenous veins. Brown et al. in 1996 described a test which used fragments of human placental arterial vessels cultured in a fibrin gel without the addition of exogenous growth factors [73]. The fetal mouse bone explant assay described by Deckers [74] involves the culture of 17-day-old fetal mouse metatarsals. As these tissues are derived from growing embryos, they contain endothelial cells from the perichondrium. This assay is suitable for testing angiogenesis inhibitors since treatment of the explants with endostatin resulted in inhibition of the tubelike structures [74].

Dorsal air sac model (DAS). Oikawa [75] first reported that the DAS model is one of the most easily prepared without requiring complex techniques as the simple implantation of a chamber ring loaded with tumor cells causes angiogenic vessel formation on the murine skin attached to the ring.

Dorsal skin fold chamber was developed in order to use the preformed tissue which is not affected by implantation procedure. Using the dorsal skin fold as chamber implantation site, those techniques were introduced during the last three decades in rats [76] and hamsters and mice [77]. Cranial window preparation was developed in an orthotopic brain tumor model by Yuan [78].

Angiomouse in the study of tumor angiogenesis. Several models previously described for angiogenesis assays by implanting tumors had big disadvantages because they lacked the microenvironment specific for each tumor type. Fluorescent protein was very useful for imaging in tumors including the formation of nascent vessels. Hoffmann et al. developed in 2002 surgical orthotopic implantation metastatic models. These models place tumors in natural microenvironments and replicate clinical tumor behavior more closely than do ectopic implantation models. The orthotopically—growing tumors, in contrast to most other models, give rise to spontaneous metastases [79].

Dual-color tumor host model. Okabe and his team developed in 1997 transgenic mice with GFP under the control of chicken beta-actin promoter and cytomegalovirus enhancer. All of the tissues from these transgenic mice, with the exception of erythrocytes and hair, have green fluorescence. Tumor cells transplanted in these mice were modified to express red fluorescent protein (RFP) [80]. Use a dual color fluorescence microscopy, Hoffmann visualized RFP-expressing tumors transplanted in the GFP-expressing transgenic mice [81]. From this model originated GFP nude mouse. The GFP nude mouse enables visualization of human tumor angiogenesis in live tissue.

2.3 Experiments "Designed" by Nature

Nature helps us. The only thing that we should do is to observe and to accept its help. Angiogenesis research has at least two such examples that might be mentioned in the present history. Both stories are related with two potent angiogenesis inhibitors. A detailed description of Ingber's accidental discovery of TNP-470 in a mold contaminating a cell culture was done by Folkman in one of his interviews. He said, "So, Don called me .... because there's something different." The cells were not dying, they were just backing away. He said there's something diffusing, something coming from the fungus that is stopping the endothelial cells (http://www.pbs.org/wgbh/nova/cancer/folkman2.html).

Donald Ingber had the presence of mind to study a curious contaminant rather than simply dispose of it. He kept in mind this experience and last year had a very fine explanation about mechanism of capillary vessels' dynamic development: "Most people think you give a growth factor and you get growth. With the same growth factor, you can get growth, you can get differentiation, you can get apoptosis—cell suicide—or you can get migration, depending on the mechanical environment." His team described for the first time the functional cross-antagonism between transcription factors that controls tissue morphogenesis, and that responds to both mechanical and chemical cues [82].

Detection of high serum levels of endostatin in patients with Down syndrome [83] explained the relative decrease in incidence of various solid tissue tumors observed in Down syndrome. Patients with this disease have an additional third copy of the *DSCR1* gene (also known as *RCAN1*) which can suppress the growth of the blood vessels that feed cancerous tumors [84]. Molecular mechanisms of angiogenesis is not yet fully understood. Down syndrome people may give us insights toward novel therapies for controlling angiogenesis in disease [85].

References

1. Willerson JT, Teaff R (1996) Egyptian contributions to cardiovascular medicine. Tex Heart Inst J 3:191–200

2. Crivellato E, Ribatti D (2006) Aristotle: the first student of angiogenesis. Leukemia 20:1209–1210

3. Hwa C, Aird WC (2007) The history of the capillary wall: doctors, discoveries, and debates. Am J Physiol Heart Circ Physiol 293: H2667–H2679

4. Gilbert SF (2008) Developmental biology, 6th edn. Wiley, New York, NY

5. Paweletz N (2001) Birth of the life sciences in The Netherlands and Belgium. Nat Rev Mol Cell Biol 2:857

6. Whitten MB (1928) A review of the technical methods of demonstrating the circulation of the heart. Arch Intern Med 42:846

7. McDonald DM, Choyke PL (2003) Imaging of angiogenesis: from microscope to clinic. Nat Med 9:713–725

8. Mendoca RJ, Mauricio VB, de Bortolli Teixeira L, Lachat JJ, Coutinho Netto J (2010) Increased vascular permeability, angiogenesis and wound healing induced by the serum of natural latex of the rubber tree Hevea brasiliensis. Phytother Res 24 (5):764–768

9. Sampaio RB, Mendoca RJ, Simioni AR, Costa RA, Siqueira RC, Correa VM, Tedesco AC, Haddad A, Coutinho Netto J, Jorge R (2010) Rabbit retinal neovascularization induced by latex angiogenic-derived fraction: an experimental model. Curr Eye Res 35:56–62

10. Schwann T, Schleiden MJ (1847) Microscopical researches into the accordance in the structure and growth of animal and plants, vol 155. The Sydenham Society, London

11. Woodward JJ (1870) Report on certain points connected with the histology of minute blood-vessels. Q J Microsc Sci 10:380–393

12. Warkany J (1981) Friedrich Daniel von Recklinghausen and his times. Adv Neurol 29:251–257

13. His W (1865) Die Häute und Höhlen des Körpers. Springer, Basel

14. Loukas M, Clarke P, Tubbs RS, Kapos T, Trotz M (2008) The His family and their contributions to cardiology. Int J Cardiol 123:75–78

15. Krogh A (1922) The anatomy and physiology of capillaries. Yale University Press, New Haven, CT, p 24

16. Florey HW, Carleton HM (1926) Rouget cells and their function. Proc R Soc Lond B 100:23–31

17. Movat HG, Fernando NVP (1964) The fine structure of the terminal vascular bed: IV. The venules and their perivascular cells (perycites, adventicial cells). Exp Mol Pathol 3:98–114

18. Allt G, Lawrenson JG (2001) Pericytes: cell biology and pathology. Cells Tissues Organs 169:1–11

19. Gospodarowicz D (1975) Purification of a fibroblast growth factor from bovine pituitary. J Biol Chem 250:2515–2520

20. Folkman J, Haudenschild CC, Zetter BR (1979) Long-term culture of capillary endothelial cells. Proc Natl Acad Sci U S A 76:5217–5221

21. Kurz H, Sandau K, Christ B (1997) On the bifurcation of blood vessels—Wilhelm Roux's doctoral thesis (Jena 1878)—a seminal work for biophysical modelling in developmental biology. Ann Anat 179:33–36

22. Ribatti D (2002) A milestone in the study of the vascular system: Wilhelm Roux's doctoral thesis on the bifurcation of blood vessels. Haematologica 87:677–678

23. Langdon SP (2004) Cancer cell culture. Methods and protocols. Humana, Totowa, NJ, pp 3–4

24. Roux W (1895) Gesammelte Abhandlungen über Entwickelungsmechanik der Organismen. Wilhelm Engelmann, Leipzig

25. Perry BN, Arbister JL (2006) The duality of angiogenesis: implications for therapy of human disease. J Invest Dermatol 126:2160–2166

26. Virchow R (1863) Die Krankhaften Geschwulste. August Hirschwald, Berlin

27. Goldman E (1907) The growth of malignant disease in man and the lower animals with special reference to the vascular system. Lancet 2:1236–1240

28. Lewis WH (1936) Malignant cells. In: The Harvey lectures, vol 31. Williams & Wilkins, Baltimore, MA, pp 214–234

29. Ide AG, Baker NH, Warren BA (1939) Vascularization of the Brown-Pearce rabbit epithelioma transplant as seen in the transparent ear chamber. Am J Roentgenol 32:891–899

30. Clark ER, Hitschler WJ, Kirby-Smith HT, Rex RO, Smith JH (1931) General observations on the ingrowth of new blood vessels into standardized chambers in the rabbit's ear, and the subsequent changes in the newly grown vessels over a period of months. Anat Rec 50:129–167

31. Clark ER, Clark EI (1939) Microscopic observations on the growth of blood capillaries in the living mammals. Am J Anat 64:251–301

32. Clark ER (1936) Growth and development of function in blood vessels and lymphatics. Ann Intern Med 9:1043–1049

33. Algire GH, Chalkley HW (1945) Vascular reactions of normal and malignant tissue in vivo. J Natl Cancer Inst 6:73–85

34. Greenblatt M, Shubik P (1968) Tumor angiogenesis: transfilter diffusion studies in the hamster by the transparent chamber technique. J Natl Cancer Inst 41:111–124

35. Folkman J (1971) Isolation of a tumor factor responsible for angiogenesis. J Exp Med 133:275–288

36. Weibel ER, Palade GE (1964) New cytoplasmic components in arterial endothelia. J Cell Biol 23:101–112

37. Wharton J (1853) Observation on the state of the blood and the blood-vessels in inflammation. Med Chir Trans 36:391–402

38. Polverini PJ (1995) The pathophysiology of angiogenesis. Crit Rev Oral Biol Med 6:230–247

39. Ausprunk DH, Folkman J (1976) Vascular injury in transplanted tissues. Fine structural changes in tumour, adult, and embryonic blood vessels. Virchows Arch B Cell Pathol 1:31–44

40. Sholley MM, Gimbrone MA Jr, Cotran RS (1977) Cellular migration and replication in endothelial regeneration: a study using irradiated endothelial cultures. Lab Invest 36:18–25

41. Figg DV, Folkman J (2008) Angiogenesis: an integrative approach from science to medicine. Springer Verlag, Berlin

42. Nowak Sliwinska P et al (2018) Consensus guidelines for the use and interpretation of angiogenesis assays. Angiogenesis 21 (3):425–532

43. Risau W, Flamme I (1995) Vasculogenesis. Annu Rev Cell Dev Biol 11:73–91

44. Risau W (1997) Mechanism of angiogenesis. Nature 386:671–674

45. Risau W (1991) Embryonic angiogenesis factors. Pharmacol Ther 51:371–376

46. Ribatti D (2010) The contribution of Roberto Montesano to the study of interactions between epithelial sheets and the surrounding extracellular matrix. Int J Dev Biol 54:1–6

47. Ribatti D (2010) A seminal work of Werner Risau in the study of the development of the vascular system. Int J Dev Biol 54:567–572

48. Ribatti D (2010) The chick embryo chorioallantoic membrane in the study of angiogenesis and metastasis. Springer, New York, NY, p 41

49. Ribatti D (2008) Judah Folkman, a pioneer in the study of angiogenesis. Angiogenesis 11:3–10

50. Folkman MJ, Long DM Jr, Becker FF (1962) Tumor growth in organ culture. Surg Forum 13:81–83

51. Folkman J, Cole P, Zimmerman S (1966) Tumor behavior in isolated perfused organs: in vitro growth and metastases of biopsy material in rabbit thyroid and canine intestinal segment. Ann Surg 164:491–502

52. Gimbrone MA Jr, Cotran RS, Folkman J (1973) Endothelial regeneration: studies with human endothelial cells in culture. Ser Haematol 6:453–455

53. Gimbrone MA Jr, Leapman SB, Cotran RS, Folkman J (1973) Tumor angiogenesis: iris neovascularization at a distance from experimental intraocular tumors. J Natl Cancer Inst 50:219–228

54. Folkman J (1990) How the field of controlled-release technology began, and its central role in the development of angiogenesis research. Biomaterials 9:615–618

55. Langer R, Folkman J (1976) Polymers for the sustained release of proteins and other macromolecules. Nature 263:797–800

56. Muthukkaruppan V, Auerbach R (1979) Angiogenesis in the mouse cornea. Science 206:1416–1418

57. Jaffe EA, Nachman RL, Becker CG, Minick CR (1973) Culture of human endothelial cells derived from umbilical veins. Identification by morphologic and immunologic criteria. J Clin Invest 52:2745–2756

58. Haudenschild CC, Cotran RS, Gimbrone MA Jr, Folkman J (1975) Fine structure of vascular endothelium in culture. Ultrastruct Res 50:22–32

59. Ingber DE, Madri JA, Folkman J (1987) Endothelial growth factors and extracellular matrix regulate DNA synthesis through modulation of cell and nuclear expansion. In Vitro Cell Dev Biol 23:387–394

60. Folkman J, Haudenschild C (1980) Angiogenesis in vitro. Nature 288:551–556

61. Carrel A (1924) Tissue culture and cell physiology. Physiol Rev 4:1

62. Maximow AA (1925) Behavior of endothelium of blood vessels in tissue culture. Anat Rec 29:369

63. Nicosia RF, Ottinetti A (1990) Growth of microvessels in serum-free matrix culture of rat aorta. A quantitative assay of angiogenesis in vitro. Lab Invest 63:115–122

64. Flamme I, Baranowski A, Risau W (1993) A new model of vasculogenesis and angiogenesis in vitro as compared with vascular growth in the avian area vasculosa. Anat Rec 237:49–57

65. Flamme I, Breier G, Risau W (1995) Expression of vascular endothelial growth factor (VEGF) and VEGF-receptor 2 (flk-1) during induction of hemangioblastic precursors and vascular differentiation in the quail embryo. Dev Biol 169:699–712

66. Krah K, Mironov V, Risau W, Flamme I (1994) Induction of vasculogenesis in quail blastodisc-derived embryoid bodies. Dev Biol 164:123–132

67. Popoff D (1894) Die Dottersack-Gefisse des Huhnes. Springer, Wiesbaden

68. Sabin FR (1920) Studies on the origin of blood-vessels and of red blood-corpuscles as seen in the living blastoderm of chicks during the second day of incubation. Contr Embryol 9:215–258

69. Rous P (1911) A sarcoma of the fowl transmissible by an agent separable from the tumor cells. J Exp Med 13:397–411

70. Murphy JB (1926) Observations on the etiology of tumors: as evidences by experiments with a chicken sarcoma. JAMA 86:1270–1271

71. Stehelin D, Varmus HE, Bishop JM (1975) Detection of nucleotide sequences associated with transformation by avian sarcoma viruses. Bibl Haematol 43:539–541

72. Auerbach R, Lewis R, Shinners B, Kubai L, Akhtar N (2003) Angiogenesis assays: a critical overview. Clin Chem 49:32–40

73. Brown KJ, Maynes SF, Bezos A, Maguire DJ, Ford MD, Parish CR (1996) A novel in vitro assay for human angiogenesis. Lab Invest 75:539–555

74. Deckers M, van der Pluijm G, Dooijewaard S, Kroon M, van Hinsbergh V, Papapoulos S, Lowik C (2001) Effect of angiogenic and anti-angiogenic compounds on the outgrowth of capillary structures from fetal mouse bone explants. Lab Invest 81:5–15

75. Oikawa T, Sasaki M, Inoue M, Shimamura M, Kubok H, Hirano S, Kumagai H, Shizuka M, Takeuchi T (1997) Effect of cytogenin, a novel microbial product, on embryonic and tumor cell-induced angiogenic responses in vivo. Anticancer Res 17:1881–1886

76. Papenfuss HD, Gross JF, Intagllieta M, Treese FA (1979) A transparent access chamber for the rat dorsal skin fold. Microvasc Res 18:311–318

77. Lehr HA, Leunig M, Menger MD, Nolte D, Messmer K (1993) Dorsal skinfold chamber technique for intravital microscopy in nude mice. Am J Pathol 143:1057–1062

78. Yuan F, Salehi HA, Boucher Y, Vasthare US, Tuma RF, Jain RK (1994) Vascular permeability and microcirculation of gliomas and mammary carcinomas transplanted in rat and mouse cranial windows. Cancer Res 54:4564–4568

79. Hoffman RM (1999) Green fluorescent protein to visualize cancer progression and metastasis. In: Michael Conn P (ed) Methods in enzymology, green fluorescent protein, vol 302. Academic, San Diego, CA, pp 20–31

80. Okabe M, Ikawa M, Kominami K, Nakanishi T, Nishimune Y (1997) 'Green mice' as a source of ubiquitous green cells. FEBS Lett 407:313–319

81. Hoffman RM (1999) Orthotopic transplant mouse models with green fluorescent protein-expressing cancer cells to visualize metastasis and angiogenesis. Cancer Metastasis Rev 17:271–277

82. Mammoto A, Connor KM, Mammoto T, Yung CW, Huh D, Aderman CM, Mostoslavsky G, Smith LE, Ingber DE (2009) A mechanosensitive transcriptional mechanism that controls angiogenesis. Nature 457:1103–1108

83. Zorick TS, Mustacchi Z, Bando SY, Zatz M, Moreira-Filho CA, Olsen B, Passos-Bueno MR (2001) High serum endostatin levels in Down syndrome: implications for improved treatment and prevention of solid tumours. Eur J Hum Genet 9:811–814

84. Baek KH, Zaslavsky A, Lynch RC, Britt C, Okada Y, Siarey RJ, Lensch MW, Park IH, Yoon SS, Minami T, Korenberg JR, Folkman J, Daley GQ, Aird WC, Galdzicki Z, Ryeom S (2009) Down's syndrome suppression of tumour growth and the role of the calcineurin inhibitor DSCR1. Nature 459:1126–1133

85. Ryeom S, Folkman J (2009) Role of endogenous angiogenesis inhibitors in Down syndrome. J Craniofac Surg 1:595–596

The Fundamental Contribution of Judah Folkman in the Setting of Angiogenesis Assays

Domenico Ribatti

Abstract

Judah Folkman (1933–2008) made seminal discoveries on the mechanisms of angiogenesis which have opened a field of investigation worldwide. This chapter summarizes the fundamental contribution of Folkman in the setting of angiogenesis assays in vivo and in vitro.

Key words Angiogenesis, Antiangiogenesis, Assays, History of medicine

1 Development of Assays to Study Angiogenesis

The study of angiogenesis in living mammalian specimens become possible in 1924 when Sandison developed a transparent chamber that could be implanted in the rabbit ear [1]. This enabled direct observation of the formation of new blood vessels in a healing wound. Judah Folkman's work started with perfused organs and then moved on to in vivo studies, first using the rat air sac model [2, 3]. As he said, "The phenomenon of angiogenesis was largely inaccessible to study. Thus throughout the middle 1970s, we were preoccupied with devising new techniques that would allow angiogenic phenomena to be quantified and permit angiogenesis to be resolved into is subcomponents." [4].

The in vivo assays of angiogenesis have enabled to make up important progress in elucidating the mechanism of action of several angiogenic factors and inhibitors. It is reasonable to reserve the term "angiogenic factor" for a substance which produces new capillary growth in an in vivo assay. A variety of animal models have been described to provide more quantitative analysis of in vivo angiogenesis and to characterize pro- and anti-angiogenic molecules. The principal qualities of the in vivo assays are their low cost, simplicity, reproducibility, and reliability which, in turn,

Domenico Ribatti (ed.), *Vascular Morphogenesis: Methods and Protocols*, Methods in Molecular Biology, vol. 2206, https://doi.org/10.1007/978-1-0716-0916-3_2, © Springer Science+Business Media, LLC, part of Springer Nature 2021

among the different in vivo assays are important determinants dictating the choice of method.

However, in vivo assays are also very sensitive to environmental factors, not readily accessible to biochemical analysis and their interpretation is frequently complicated by the fact that the experimental conditions favor inflammation, and that under these conditions the angiogenic response is elicited indirectly, at least in part, through the activation of inflammatory or other non-endothelial cells.

As pointed out by Auerbach et al. [5], "Perhaps the most consistent limitation to progress in angiogenesis research has been the availability of simple, reliable, reproducible, quantitative assays of the angiogenic response." On the basis of these limitations, ideally two different assays should be performed in parallel to confirm the angiogenic or anti-angiogenic activities of test substances.

As pointed out by Zetter in 2008, "At the time that Folkman began his studies on the relationship between angiogenesis and tumorigenesis, there were a few in vivo bioassays that had provided the majority of the earlier data in the field (...) In the 1960s and 1970s, Folkman and his colleagues developed or perfected nearly all the in vivo and in vitro assays that are used in the field today." [6]

The classical assays for studying angiogenesis in vivo include the hamster cheek pouch, the rabbit ear chamber, the dorsal skin and air sac, the iris and avascular cornea of rodent eye, and the chick embryo chorioallantoic membrane (CAM). For a reliable and reproducible assessment of angiogenesis for all of the assays, validation procedures and quality control protocols are mandatory.

2   Tumor Growth in Isolated Perfused Organs

In a series of experiments performed in the 1960s when he was enlisted in the US National Naval Medical Center, Folkman found that implanted tumor cells did not grow in various organs perfused with blood substitutes preparations [7, 8]. However, the same tumor cells did grow when implanted in mice (Fig. 1).

As Folkman pointed out, "When the implants failed to grow beyond 1.5–2 mm, our first thought was that the area of central necrosis was in some way toxic to the peripheral cell layer. However, when the tumors were removed from these isolated organs and transplanted back to the host animals, large tumors grew that killed the host. When histological sections of the in vitro implants were compared to the in vivo implants, there was a striking difference. The in vitro implants were avascular, whereas the in vivo implants were large and well vascularized." [10].

In collaboration with Michael Gimbrone, Folkman demonstrated that platelet-rich medium was necessary for the preservation of isolated organs and short-term tumor cell growth after

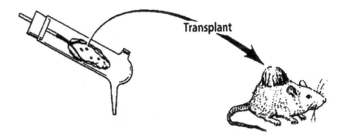

Fig. 1 When a no expanding tumor is transplanted to a syngeneic mouse, it grows more than 1000 times its initial volume in a perfused organ (Reproduced from [9])

implantation in perfused organs [11, 12]. A majority of these tumor implantations grew to a size of 3–4 mm in diameter for up to 4 weeks and Folkman concluded that tumor growth in isolated perfused organs was restricted to a microscopic size in the absence of neovascularization.

3 The Rabbit Cornea

The use of the rabbit cornea as a normally avascular tissue that could support the growth of new vessels when stimulated by a tumor was pioneered by Folkman and Gimbrone (Fig. 2) [11, 14]. As Folkman said. "Michael Gimbrone, a post-doctoral fellow, and I implanted tumors of approximately 0.5 mm^3 into the stromal layers of the rabbit cornea at a distance of up to 2 mm from the limbal edge. New capillary blood vessels grew from the limbus, invaded the stroma of the avascular corneas, and reached the edge of the tumor over a period of approximately 8–10 days (Fig. 3). When tumors were implanted beyond 3 mm from the limbus (or in the center of the rabbit cornea), no neovascularization was observed" [15].

The experimental design was to make a hollow pocket in the cornea and implant the tumor in the pocket. Blood vessel sprouts would then move across the previously avascular cornea toward the tumor. Robert Langer and Folkman [16] developed the use of polymeric systems that could deliver proteins continuously over an extended time period using this assay.

As Folkman pointed out, "Langer dissolved the polymer poly-hydroxyethylmethacrylate (polyhema), into alcohol and added lyophilized protein. When the solvent was evaporated, the protein remained trapped in a rubbery polymeric pellet. When the pellet was implanted into the cornea, water diffused into the pellet. This caused the formation of microchannels around the protein. Protein diffused out from these channels at zero-order kinetics for weeks to months" [15]. As Zetter said, "This assay allowed the first analysis

Fig. 2 Judah Folkman in collaboration with his technician photographing a rabbit eye (Reproduced from [13])

of fractionated angiogenic factors that had been isolated from tumour cells. It also opened up an entirely new field of slow release delivery of macromolecules." [6].

Ziche and Gullino [17] by using the rabbit cornea demonstrated that angiogenesis is a marker of neoplastic transformation. Normal mouse diploid fibroblasts were carried in culture. At each passage, the cells were tested for angiogenic activity in the rabbit eye, and for tumorigenicity by reimplantation into the mouse strain that donates the fibroblasts. Angiogenesis first appeared at the fifth passage, while tumorigenicity did not occur until the 15th passage. This assay was subsequently refined by introducing the tumor cells into the stroma of the cornea, and then by substituting slow-release pellets containing known quantities of semipurified angiogenic growth factors from tumor cells. Nonspecific angiogenic stimuli such as endotoxin were used initially, but were replaced with specific growth factors such as fibroblast growth factor-2 (FGF-2) and vascular endothelial growth factor (VEGF).

After initial looping of limbal capillaries, loops extend into the cornea and vascular sprouts appear at the apices of the loops, and finally a vascular network develop toward the tumor implant.

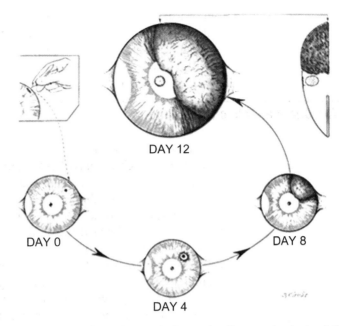

Fig. 3 The patterns of development of two simultaneous implants of Brown-Pearce tumor in the rabbit's eye. The anterior chamber implant remains avascular, while the iris implant vascularizes and grows progressively (Reproduced from [9])

The rabbit size (2–3 kg) lets an easy manipulation of both the whole animal and the eye to be easily extruded from its location and to be surgically manipulated. Sodium pentobarbital is used to anaesthetize animals and under aseptic conditions a micropocket (1.5 × 3 mm) is produced using a pliable iris spatula 1.5 mm in width in the lower half of the cornea. A small amount of the aqueous humor can be drained from the anterior chamber when reduced corneal tension is required. For a corneal transplant, a 1.5 mm incision is made just off center of the corneal dome to a depth of about one-half the thickness of the cornea. Substances used within the pocket have included tumor tissues, tumor cells, tumor cell extracts, other tissues and cells, concentrated conditioned medium, purified recombinant cytokines and growth factors. The implant is located at 2.5–3 mm from the limbus to avoid false positives due to mechanical procedure and to allow the diffusion of test substances in the tissue, with the formation of a gradient for the endothelial cells of the limbal vessels. A disadvantage is that tumors other than those of rabbit origin may induce an immune response once they are vascularized and subsequent inflammatory angiogenesis may superimpose to primary angiogenic response.

Several angiogenic factors in combination can be simultaneously implanted into the corneal tissue and synergistic or antagonistic angiogenic response can be analyzed. Moreover, inhibition of corneal neovascularization can be achieved by systemic administration of angiogenesis inhibitors. One of the first inhibitors of angiogenesis studied by means of this assay was an extract of cartilage [18].

4 The Chorioallantoic Membrane

Folkman used also another assay to study angiogenesis in vivo, namely, the chick embryo CAM (Fig. 4) [19, 20]. The vessels of the CAM grow rapidly until day 11 of incubation [21]. The unknown test material was dried in 5 μl H_2O on a sterile plastic coverslip and this disk was the placed on the CAM of a nine or ten-day chick embryo through a window previously made in the shell. The test substance can be also dissolved in 0.5% methyl cellulose, which was then dried to make disks of about 2 mm diameter [22]. The presence of angiogenesis was determined 48 h later by observing new capillaries converging on the disc.

Tumors or tumor fractions implanted on day 9 induced an angiogenic response within 48–72 h, and this response can be recognized under a stereomicroscope as new capillaries converging on the implant [23]. These authors, working in the Folkman's laboratory, using implants of fresh Walker 256 carcinoma, described the onset of tumor vascularization in the CAM from day 5 to 16 of incubation. Chick capillaries proliferated in the vicinity of the tumor graft about 24 h after implantation, but capillary sprouts did not penetrate the graft until approximately 72 h later. During the avascular interval, tumor diameter did not

Fig. 4 A port trait of J. Folkman in the early 1970s with chicken eggs (Reproduced from [9])

exceed 1 mm, but grew rapidly during the first 24 h following capillaries penetration. Tumors implanted on the CAM of older embryos grew at a slower rate in parallel with the reduced rate of endothelial growth.

The CAM bioassay is carried out in an egg in which a window is prepared by removing a piece of the eggshell. Folkman proposed an alternative method by incubating the embryos in a petri dish by totally removing the shell [19]. In these shell-less embryos, Folkman found that the yolk sac vessels could be used on day 4 and the CAM on day 6 to test either pro-angiogenic and anti-angiogenic molecules [22, 24]. Shell-less culture of avian embryos facilitates experimental access and continuous observations of the growing embryos almost to the term of hatching. The embryo and its extraembryonic membranes may be transferred to a petri dish on day 3 or 4 of incubation and CAM develops at the top as a flat membrane and reaches the edge of the dish to provide a two-dimensional monolayer onto which multiple grafts can be placed [19]. This system has several advantages as the accessibility of the embryo is greatly improved outside of the shell. Shell-less culture is much more amenable to live imaging than in ovo techniques. However, long-term viability is often lower in shell-less cultures and great attention must be paid to preventing the embryo from drying out. In the original description of embryos cultured in petri dishes, there was a 50% loss in the first 3 days after cracking due to the frequent rupture of the yolk membrane, with 80% of those which survive to day 7 until day 16 [19]. The ex ovo method is preferred to the in vivo method because it allows for the quantification of the response over a wider area of the CAM.

Another method has been proposed by Folkman and collaborators [25]: the testing substance is placed into a collagen gel between two parallel nylon meshes which align the capillaries for counting. The resulting "sandwich" is then placed upon the CAM on day 8 of incubation. This method was based on the vertical growth of new capillaries into the collagen gel containing FGF2 and sucralfate or tumor cells (Fig. 5). This assay does not depend on flow, nor does it require injection of dyes, measurements of DNA synthesis, and image analysis. A major advantage of this method is that it does not require histological sections, thus facilitating the screening of o large number of compounds. The results are expressed as a percentage of the squares in the top mesh that contained blood vessels.

The CAM has been established as an experimental system for research in tumor biology, as a test system for tumor chemosensitivity, for the study of tumor invasion and metastasis and of neovascularization of heterologous normal and neoplastic implants.

The main limitation of CAM is represented by no specific inflammatory reactions which may develop as a result of grafting, and in turn induce a secondary vasoproliferative response eventually making it difficult to quantify the primary response that is being

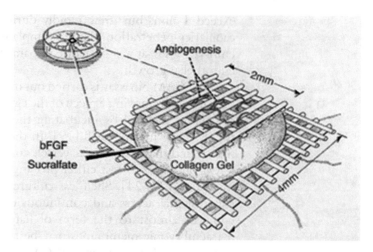

Fig. 5 FGF2 and sucralfate are mixed in a collagen gel embedded between two pieces of mesh and placed on the CAM. The results are expressed as the percentage of the squares in the top mesh that contain blood vessels (Reproduced from [9])

investigated. Inflammatory angiogenesis per se in which infiltrating macrophages or other leukocytes may be the source of angiogenic factors cannot be distinguished from direct angiogenic activity of the test material without detailed histological study and multiple positive and negative controls. In this connection, a study of histological CAM section would help detecting the possible presence of a perivascular inflammatory infiltrate together with a hyperplastic reaction, if any, of the chorionic epithelium. However, the possibilities of causing nonspecific inflammatory response are much lower when the test material is grafted as soon as CAM begins to develop since then the host's immune system is relatively immature. This problem may be overcome by using the yolk sac vessels of the 4-day chick embryo because this system has a markedly reduced inflammatory and immune response.

5   Culture of Endothelial Cells

Culture of endothelial cells was first reported in 1973–74, independently by Folkman's laboratory [14] and by Eric Jaffe at Cornell University, Ithaca, New York, USA [26]. As Folkman remembered: "Before 1970, no one had been able to culture and passage vascular endothelial cells of any kind. In 1972, Jaffe and his colleagues succeeded with human umbilical vein endothelial cells. Gimbrone and I immediately employed this method in our laboratory. We grew the endothelial cells to confluence, at which point they became quiescent (Fig. 6). With great excitements we added our most potent angiogenic extracts, hoping to see the same burst of

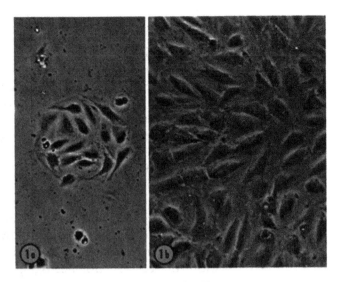

Fig. 6 Capillary endothelial cells in culture viewed by phase contrast microscopy (Reproduced from [9])

DNA synthesis that occurs when growth factors are added to confluent cultures of 3T3 fibroblasts. Unfortunately, nothing happened." [4].

As Auerbach and Auerbach [27] pointed out, "These cultures, obtained from human umbilical veins, and subsequent cultures obtained from other large vessels, such as bovine aorta, pulmonary vein and pulmonary artery, were prepared by mild digestion of the internal layer of the vessels by ligation of a segment of the blood vessel, introduction of an enzyme solution into the sealed-off vessel segment, brief incubation, and finally, the elution of the loosened internal lining that was composed of vascular endothelial cells."

As Folkman said, "At least three possible explanations the failure of umbilical vein endothelial cells to respond to angiogenic factors that had consistently stimulated capillary proliferation in vivo: (1) Confluent cultures of endothelial cells could be refractory to any further growth stimulation (in contrast to 3T3 fibroblasts). (2) An intermediate cell was missing in vitro: perhaps mast cells or macrophages were needed to process an angiogenic factor before endothelial cells could respond. (3) Endothelial cells from large vessels might be the wrong type of cells. Perhaps capillary endothelial cells would be more responsive. (...) The third hypothesis led to five frustrating years of attempts to culture capillary endothelial cells" [4].

After a series of unsuccessful experiments, Folkman and his collaborators Haudenschild and Zetter, were the first to achieve long-term culture of capillary endothelial cells from the bovine adrenal cortex [28], and induced capillary endothelial cells to form three-dimensional networks in vitro that had some properties of capillary networks in vivo [29]. By plating a dilute suspension of the initial cell isolates, they were able to spot individual small islands of endothelial cells distinct from other islands of fibroblasts and pericytes. By manually removing the no endothelial cells each day, they obtained a culture dish enriched for endothelial cells. In such dishes, colonies of endothelial cells could be encircled and plucked for subculture.

As Folkman pointed out, "When capillary endothelial cells reach confluence they form tubes that resemble capillaries in vivo. The first tubes arise from a cylindrical vacuole that appears within a single endothelial cell. Contiguous cells develop similar vacuoles, which by connecting one cell to another form a long tube. Branches begin to appear after four of five cells are connected and within five to ten days an entire network of tubes may replace the dense regions of the culture. Branch formation can be observed within one cell. Electron micrographs show that the tubes represent continuous lumina bordered by cell cytoplasm and membranes" [30].

The ultrastructure of cultured endothelial cells is identical to that of endothelial cells in situ [31]. Functional studies have shown that when endothelial cells are shown in appropriate culture conditions, they form intercellular junctions resembling those of endothelium in terms of their permeability properties in vivo.

References

1. Sandison JC (1924) A new method or the microscopic study of living growing tissues by introduction of a transparent chamber in the rabbit's ear. Anat Rec 28:281–287

2. Folkman MJ, Long DM, Becker FF (1963) Growth and metastasis of tumor in organ culture. Cancer 16:453–467

3. Folkman J, Merler E, Abernathy C et al (1971) Isolation of a tumor fraction responsible for angiogenesis. J Exp Med 133:275–288

4. Folkman J (1985) Toward an understanding of angiogenesis: search and discovery. Perspect Biol Med 29:10–36

5. Auerbach R, Auerbach W, Polakowski J (1991) Assays for angiogenesis: a review. Pharmacol Ther 51:1–11

6. Zetter BR (2008) The scientific contribution of M. Judah Folkman to cancer research. Nat Rev Cancer 8:647–654

7. Folkman J, Cole P, Zimmerman S (1966) Tumor behavior in isolated perfused organs: in vitro growth and metastases of biopsy material in rabbit thyroid and canine intestinal segment. Ann Surg 164:491–502

8. Folkman J, Winsey S, Cole P et al (1968) Isolated perfusion of thymus. Exp Cell Res 53:205–214

9. Ribatti D (2018) Judah Folkman. A biography. Springer International Publishing AG, Cham

10. Folkman J, Cotran R (1976) Relation of vascular proliferation to tumor growth. Int Rev Exp Pathol 16:207–248

11. Folkman J, Gimbrone MA Jr (1975) Perfusion of the thyroid gland. Methods Enzymol 39:359–364

12. Gimbrone MA Jr, Aster RH, Cotran RS et al (1969) Preservation of vascular integrity in organs perfused in vitro with a platelet-rich medium. Nature 222:33–36

13. Cimpean AM, Ribatti D, Raica M (2011) A brief history of angiogenesis assays. Int J Dev Biol 55:377–382

14. Gimbrone MA Jr, Cotran RS, Folkman J (1974) Tumor growth and neovascularization: an experimental model using rabbit cornea. J Natl Cancer Inst 52:413–427

15. Folkman J (2008) Tumor angiogenesis: from bench to bedside. In: Marmé D, Fusenig N (eds) Tumor angiogenesis. Basic mechanisms and cancer therapy. Springer, New York, NY, pp 3–28

16. Langer R, Folkman J (1976) Polymers for the sustained release of proteins and other macromolecules. Nature 263:797–800

17. Ziche M, Gullino PM (1982) Angiogenesis and neoplastic progression in vitro. J Natl Cancer Inst 69:483–487

18. Brem H, Folkman J (1975) Inhibition of tumor angiogenesis mediated by cartilage. J Exp Med 141:427–439

19. Auerbach R, Kubai L, Knighton D et al (1974) A simple procedure for the long-term cultivation of chicken embryos. Dev Biol 41:391–394

20. Ausprunk DH, Knighton DR, Folkman J (1975) Vascularization of normal and neoplastic tissues grafted to the chick chorioallantois. Role of host and preexisting graft blood vessels. Am J Pathol 79:597–618

21. Ausprunk DH, Folkman J (1977) Migration and proliferation of endothelial cells in preformed and newly formed blood vessels during tumor angiogenesis. Microvasc Res 14:53–65

22. Taylor S, Folkman J (1982) Protamine is an inhibitor of angiogenesis. Nature 297:307–312

23. Knighton D, Ausprunk D, Tapper D et al (1977) Avascular and vascular phases of tumour growth in the chick embryo. Br J Cancer 35:347–356

24. Folkman J, Langer R, Linhardt R et al (1983) Angiogenesis inhibition and tumor regression caused by heparin or a heparin fragment in the presence of cortisone. Science 221:719–725

25. Nguyen M, Shing Y, Folkman J (1994) Quantitation of angiogenesis and antiangiogenesis in the chick embryo chorioallantoic membrane. Microvasc Res 47:31–40

26. Jaffe EA, Nachman RL, Becker CG et al (1973) Culture of human endothelial cells derived from umbilical veins. Identification by morphologic and immunologic criteria. J Clin Invest 52:2745–2756

27. Auerbach R, Auerbach W (2001) Assays to study angiogenesis. In: Voest EE, D'Amore PA (eds) Tumor angiogenesis and microcirculation. Marcel Dekker Inc, Basel, pp 91–102

28. Folkman J, Haudenschild C, Zetter BR (1979) Long-term culture of capillary endothelial cells. Proc Natl Acad Sci U S A 76:5217–5221

29. Folkman J, Haudenschild C (1980) Angiogenesis in vitro. Nature 288:551–556

30. Folkman J (1982) Angiogenesis: initiation and control. Ann N Y Acad Sci 401:212–226

31. Haudenschild CC (1984) Morphology of vascular endothelial cells in culture. In: Jaffe EA (ed) Biology of endothelial cells. Martinus Nijhoff Publishers, Boston, MA, pp 129–140

Chapter 3

Identification of Endothelial Cells and Their Progenitors

Ellen Go and Mervin C. Yoder

Abstract

Blood vessel formation is a key feature in physiologic and pathologic processes. Once considered a homogeneous cell population that functions as a passive physical barrier between blood and tissue, endothelial cells (ECs) are now recognized to be quite "heterogeneous." While numerous attempts to enhance endothelial repair and replacement have been attempted using so called "endothelial progenitor cells" it is now clear that a better understanding of the origin, location, and activation of stem and progenitor cells of the resident vascular endothelium is required before attempting exogenous cell therapy approaches. This chapter provides an overview for performance of single-cell clonogenic studies of human umbilical cord blood circulating endothelial colony-forming cells (ECFC) that represent distinct precursors for the endothelial lineage with vessel forming potential.

Key words Endothelial colony-forming cells, Colony-forming assay, Cell proliferation, Endothelial outgrowth cells, Blood outgrowth endothelial cell, Single-cell assay

1 Introduction

The endothelium is a single layer of cells that forms the inner lining of the entire cardiovascular system. It is a dynamic, heterogeneous organ of arterial, venous, and capillary vessels that possess vital secretory, synthetic, metabolic, and immunologic functions [1–4]. There is marked phenotypic variation between endothelial cells in different parts of the vascular system, expressing different surface antigens, receptors, and generate different physiologic responses that are organ specific [5, 6]. Following the development of single-cell transcriptomic analysis, tissue-specific endothelial heterogeneity and contributions to blood vessel formation and function are now becoming understood at the molecular level [4, 7–10]. An additional emerging theme is that resident vascular endothelial cells possess a subset of cells that display stem- and progenitor-like properties that contribute to physiologic and pathologic events [11–16]. These cells play a major role in endothelial repair and regeneration in response to injury. Reported methods to isolate and identify the resident vascular endothelial stem cells

Domenico Ribatti (ed.), *Vascular Morphogenesis: Methods and Protocols*, Methods in Molecular Biology, vol. 2206,
https://doi.org/10.1007/978-1-0716-0916-3_3, © Springer Science+Business Media, LLC, part of Springer Nature 2021

include use of flow cytometry and cell sorting, in vitro 3D lumenized capillary-like structure formation, in vivo vessel forming assays, rescue of vasculature following induced ischemic and hypoxic tissue injury, and in vitro clonogenic assays (reviewed in [12, 17, 18]). While most of these tools have been used to identify murine resident vascular endothelial stem cells, the field continues to wait on the prospective identification of a cell with similar properties in the human vascular system. This chapter will provide a detailed method for identifying human umbilical cord blood ECFC that display clonogenic activity and describe how to enumerate them.

The term endothelial colony-forming cell (ECFC) was first coined in 2004 to discriminate those circulating blood endothelial cells that displayed proliferative potential when plated at a single-cell level compared to similarly plated and cultured endothelial cells that merely attached but never divided [19]. ECFC were found to be more frequent in umbilical cord blood than in adult peripheral blood [19]. ECFC were subsequently identified from cultured single cells derived from umbilical artery and veins, as well as human adult aorta [20]. Whether isolated from the circulation or as a resident endothelial cell, ECFC that emerged as clones of proliferating cells displayed a unique hierarchy of proliferative potentials with the most proliferative cells (clones forming colonies of >2000 cells within 10 days of culture initiation that displayed self-renewal potential) possessing the highest level of telomerase activity [19, 20]. Numerous subsequent studies have identified various markers, drugs, growth factors, and disease states that differentially impact the proliferative potential of the ECFC [21–24]. Of interest, ECFC have been shown to possess robust in vivo vessel forming activity and have been reported to successfully rescue blood flow in a variety of preclinical animal models of human diseases marked by hypoperfusion, ischemia, and hypoxia (reviewed in [12, 25–27]). The growing translational interest has led to the publication of several position papers calling for standardization of methods to quantify and culture ECFC derived from peripheral blood [17, 28].

2   Materials

Prepare all medium and solutions using deionized ultrapure water. Store all reagents according to the manufacturer's instructions.

2.1   Reagents

1. 4′,6-diamidino-2-phenylindole (DAPI).

2. Antibiotic–antimycotic (PSA; 100×): penicillin (10,000 U/ml)–streptomycin (10,000 µg/ml)–amphotericin (25 µg/ml).

3. Dimethyl sulfoxide (DMSO).

4. Defined fetal bovine serum (FBS; Hyclone).

5. Cryopreservation medium: Mix FBS and DMSO at 9:1 dilution under sterile conditions. Use only freshly prepared freezing medium.

6. Endothelial Cell Growth Medium-2 BulletKit™ (EGM™-2; Lonza): contains endothelial basal medium 500 ml and supplements: Hydrocortisone 0.20 ml, hFGF-B 2.00 ml, VEGF 0.50 ml, R3-IGF-1 0.50 ml, ascorbic acid 0.50 ml, hEGF 0.50 ml, GA-1000 0.50 ml, heparin 0.50 ml.

7. Complete EGM-2 medium (cEGM-2): EGM™-2 medium is supplemented with 10% FBS and 1.5% antibiotic–antimycotic solution. Culture medium can be stored for up to 4 weeks at 4 °C.

8. Ethanol, 70%.

9. Ficoll-Paque (Fisher Scientific).

10. Glacial acetic acid, 17.4 N.

11. Heparin.

12. Paraformaldehyde, 4%.

13. Phosphate buffered saline (PBS), Add 900 ml water to a 1 l beaker. Add 8 g sodium chloride (137 mM NaCl, final concentration). Add 0.2 g potassium chloride (2.7 mM KCl, final concentration). Add 1.44 g disodium hydrogen phosphate (10 mM Na_2HPO_4, final concentration). Add 0.24 g monopotassium phosphate (: 1.8 mM KH_2PO_4, final concentration). Adjust the pH to 7.4 with HCl, and then transfer the solution from the beaker to a graduated cylinder. Adjust the volume to 1 l with water. Aliquot the PBS into 500 ml bottles and autoclave.

14. Trypan blue solution, 0.4% wt/vol.

15. TrypLE Express Enzyme (Gibco).

16. Turk blood diluting fluid.

17. Type I rat tail collagen (BD Bioscience): Dilute 0.575 ml of glacial acetic acid in 493.5 ml of water, and sterile filer using a 0.22-μm vacuum filtration system. Then, add Type I rat tail collagen to dilute acetic acid bring to a final concentration of 50 μg/ml. Store solution for up to 4 weeks at 4 °C. Add 1 ml rat tail collagen type I solution to each well of a 6-well tissue culture plate (5 ml into 75 cm^2 flask) and incubate for at least 30 min at 37 °C with 5% CO_2. Remove collagen solution, then wash plate once with PBS (for Subheading 3.1). Add 100 ml rat tail collagen type I solution to each well of a 96-well plate and incubate for at least 30 min at 37 °C with 5% CO_2. Remove collagen solution, then wash plate with PBS (for Subheading 3.6).

18. Cells.

2.2 Equipment	1. 37 °C water bath.
	2. 0.22-µm vacuum filtration system.
	3. 6-well culture plates.
	4. 75 cm^2 culture flasks.
	5. 96-well tissue culture plates.
	6. Cell culture centrifuge.
	7. Cell Freezing Device (Nalgene Mr. Frosty Freezing Container).
	8. Conical centrifuge tubes, 50 ml.
	9. Cryogenic vial, 2 ml.
	10. Flow sorting machine (Becton Dickinson FACSAria Sorter).
	11. Fluorescence microscope (Zeiss Axiovert 25 CFL inverted microscope with a 10× CP-ACHROMAT/0.12 NA objective).
	12. Freezer, −80 °C.
	13. Hemocytometer.
	14. Humidified 37 °C incubator with 5% carbon dioxide (CO_2).
	15. Multichannel pipette.
	16. Pipettes.
	17. Tissue culture biohazard safety hood.

3   Methods

Prepare all medium and solutions using deionized ultrapure water and analytical grade reagents. Work in aseptic BSL2 conditions (since using primary human tissues to isolate ECFC) under the cell culture hood at all time.

3.1 ECFC Isolation and Expansion

1. Collect cord blood (CB) with 60 ml syringe containing heparin (10 U heparin/ml blood) and transport to the laboratory at room temperature for processing (*see* **Note 1**). Upon return to the laboratory, dilute CB 1:1 in sterile PBS, then pipet several times to mix.

2. In a separate 50-ml conical centrifuge tube, overlay 20 ml of the CB–PBS mixture onto 10 ml of Ficoll-Paque, being careful to avoid disrupting the layers (*see* **Note 2**). Centrifuge at 500 × *g* for 30 min at room temperature, without engaging deceleration brake setting.

3. Aspirate the buffy coat layer of low-density mononuclear cells (MNCs) at the interface between the Ficoll-Paque and diluted plasma and transfer to a new 50-ml conical tube (we recommended combining two buffy coat layers in each tube). Then add PBS to make total volume of 45 ml.

4. Centrifuge at $300 \times g$ for 10 min at room temperature, with high deceleration brake setting. Carefully remove the supernatant and combine cell pellets into a single tube.

5. Resuspend the cells in cEGM-2 medium and count using Turk blood diluting fluid at 1:10 dilution. Plate MNC at 10,000 cells/cm^2 in 4 ml of cEGM-2 media into each well of collagen I coated 6-well tissue culture plates and place inside a 37 °C, 5% CO$_2$ humidified incubator.

6. After 24 h (day 1), slowly and carefully remove the spent medium with a pipette so as not to visually disturb the loosely adherent cells (these cells contain the ECFC), and add 4 ml of cEGM-2 to the well before returning the plate to the incubator. Repeat this step of medium change daily until day 7, then every other day thereafter until day 14. Colonies typically appear between days 7–10.

7. On day 14, remove the spent media and wash the plate with PBS. Remove PBS and harvest colonies by adding 1 ml of TrypLE express to each well. Incubate for 3 min at 37 °C until cells begin to detach. After 3 min, gently tap the sides and bottom of the plate to remove remaining adherent cells of the colonies. Collect the colonies into a new 50-ml conical tube containing 1 ml of FBS and pipet up and down to generate a single-cell suspension. Centrifuge at $300 \times g$ for 5 min to pellet the ECFC.

8. Remove supernatant and resuspend cell pellet in cEGM-2-medium. Plate the cells onto collagen I–coated 75 cm^2 flask at a density of 10,000 cells/cm^2 and place inside a 37 °C, 5% CO$_2$ humidified incubator for expansion. Change media every other day until cells approach 80–90% confluency (Passage 1) before subculturing again.

9. We recommend performing routine phenotypic and functional characterization of ECFC using microscopy, immunostaining or FACS methods, and 3D vascular network formation in vitro. Bona fide ECFC form a monolayer of cells with cobblestone morphology (Fig. 1a), express EC-specific markers like CD31 and CD144, and do not express hematopoietic specific markers like CD45 and CD14. They are identified as CD144+CD31 +CD34+CD14−CD45− cells that are capable of forming in vitro lumenized tubes in three-dimensional collagen gels (Fig. 1b) [28, 29].

3.2 ECFC Cryopreservation

1. Harvest ECFC in the log phase of growth which is equivalent to 80–90% confluency. Count the number of cells to be cryopreserved using the hemocytometer.

2. Resuspend the cells in freezing medium to a concentration of 1×10^6 cells/ml.

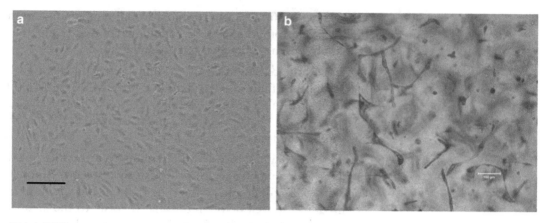

Fig. 1 ECFC properties. (**a**) Cobblestone morphology of ECFC with 80–90% confluency. (**b**) Formation of tubes with lumen in 3D collagen matrix. *ECFC* endothelial colony forming cell. Scale bar, 100 μm

3. Aliquot 1 ml of cells into properly labeled cryogenic vials and place vials in Mr. Frosty system to ensure a gradual cooling rate of between −1 and −3 °C/min. Place in −80 °C freezer overnight and the next morning, transfer frozen cryogenic vials to a storage freezer box at −80 °C for holding (no more than 3 months), or to storage boxes in a liquid nitrogen vessel for long term storage.

3.3 Thawing ECFC

1. Thaw early passage (P) ECFC (P2–3) by placing the cryogenic vial at room temperature for 2–3 min. When tubes begin to thaw (tubes will frost and frozen pellet will move upon vial inversion), place them in 37 °C water bath with gentle tapping until there is just a small amount of unfrozen material left (*see* **Note 3**).

2. Spray tube with 70% ethanol and transfer to tissue culture hood.

3. Slowly add 1 ml FBS to the opened cryogenic vial, gently aspirate the contents, and transfer the cells to a 50 ml conical tube with serum-containing PBS or media (the final concentration is 10% FBS). Centrifuge at $300 \times g$ and resuspend cell pellet in 10 ml of cEGM-2 media.

3.4 ECFC Culture and Expansion

1. Wash the type I rat tail collagen–coated flasks twice with PBS immediately before cell plating (*see* **Note 4**).

2. Transfer cell mixture to the flask and place it in a humidified 37 °C incubator with 5% CO_2. Change media every 3 days until cells have reached 70–80% confluency.

3. Expand the cells by seeding about 10,000 cells/cm^2 onto a collagen I–precoated tissue culture surfaces in cEGM-2 media with media change every three days until cells approach 70–80% confluency before subculturing again (*see* **Note 5**).

3.5 Single-Cell Preparation

1. Detach the cells by adding TrypLE express (5 ml in T75 flask) and then placing inside the incubator for 3–5 min until cells begin to detach.

2. Transfer cells to conical tubes. Wash the flask two times with 5–10 ml cEGM-2 medium to collect remaining cells and transfer the washings into the tube. Centrifuge at $300 \times g$ for 5 min.

3. Remove the supernatant and resuspend the cell pellet in 1 ml cEGM-2 medium. Pipet up and down to generate single-cell suspension.

4. Obtain an aliquot and determine cell count and viability using a hemocytometer and trypan blue exclusion (*see* **Note 6**).

3.6 Single-Cell Clonogenic Assay

1. Wash the Type I rat tail collagen–coated wells twice with PBS immediately before cell plating.

2. Use FACSAria (Becton Dickinson) or other comparable sorter with low flow rate of 20 cells/second to sort one single ECFC per well on a 96-well plate based on FSC, SSC, and DAPI stain. Adjust the final volume to 200 µl per well with complete EGM-2 medium. Incubate the plate in a 37 °C incubator in room air with 5% CO_2 overnight (*see* **Note 7**).

3. The next day, examine the plate using a microscope and take note of wells that have only one cell deposited in it. The media can be replaced with solutions containing a test substance to be tested (example, if assessing for the effect of drugs on ECFC proliferation).

4. Incubate for 13 days with two media changes performed on day 5 and day 10. Gently aspirate the spent medium and add 200 µl fresh medium slowly to prevent cells from detaching (*see* **Note 8**).

5. On day 14 of culture, aspirate the media and wash once with 200 µl of PBS. Fix cells by adding 100 µl of 4% paraformaldehyde and let sit for 15 min at room temperature, then wash three times with PBS before adding 100 µl of DAPI solution at 1 µg/ml concentration (*see* **Notes 9** and **10**). Incubate with DAPI for 10–15 min and wash twice with PBS (*see* **Note 11**).

6. Examine the wells for cell expansion and quantitate the number of endothelial cells in each well under a fluorescent microscope by counting DAPI stained cells. All wells with at least two endothelial cells are considered positive and scored.

7. Report scores by categorizing them as endothelial cell clusters if there are 2–50 cells, low proliferative potential (LPP)-ECFCs if there are 51–2000 cells, or high proliferative potential (HPP)-ECFCs if there are 2001 or more cells [19]. Schematic representation in Fig. 2.

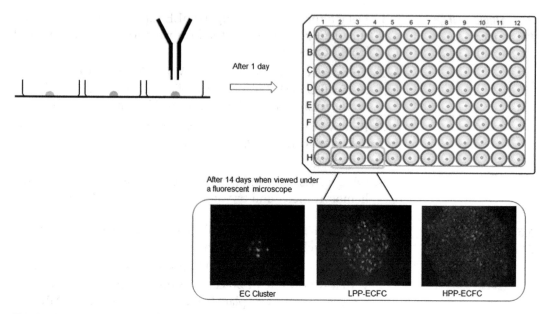

Fig. 2 Schematic of single-cell assay using cord-blood derived ECFC. Green fluorescence protein (GFP)-expressing ECFC viewed under a fluorescence microscope after 14 days of culture. From a single cell, ECFC can give rise to EC clusters (2–50 cells) or colonies (51–2000 cells or 2001 or more cells per colony) in 96-well plates. Scale bar, 100 μm. *EC* endothelial cell, *ECFC* endothelial colony forming cell, *LPP-ECFC* low proliferative potential ECFC, *HPP-ECFC* high proliferative potential ECFC [19]

4 Notes

1. Heparin is important for preventing the blood from coagulating. Heparin stock solution should be kept at 4 °C and new stock solution should be made every 4 weeks. We make heparin stock solution by dissolving heparin salt powder from Sigma-Aldrich with water to make 10,000 U/ml and filtering the solution through a 0.22 μm filter membrane. Commercially available ready-made heparin stock solutions can also be used.

2. Draw up Ficoll-Paque into a 20-ml syringe and attach a 20 G needle or mixing cannula. Place the tip of the mixing cannula at the bottom of the tube of diluted blood and carefully underlay Ficoll-Paque. The diluted blood will rise up the tube as the Ficoll-Paque is added. Proper underlays will display a clear crisp line of demarcation between the diluted blood and the Ficoll-Paque.

3. Early passage cells contain a higher cell purity, well-preserved biologic function, and phenotypic ECFC markers than later passage cells [30].

4. Type 1 rat tail collagen provides a surface bed for cell growth and expansion that closely resembles the interstitial extracellular matrix protein in humans [31]. Others have used fibronectin which is equally efficient [32]. Gelatin has also been used with success to grow ECFC [33].

5. Cell density is critical as confluency can alter the proliferative properties of the cell.

6. Alternatively, single cells can be plated manually using a multichannel pipette but accurate dispensing is extremely important. In this method, one dilutes the cell population to a concentration of 0.3 cells/100 μl complete EGM2 media and dispense cells with mixing of the cell suspension with each pipette discharge and reload. Then add an additional 100 μl of complete EGM-2 media as in Subheading 3.8. With this method, there is a higher chance of depositing no cells with approximately 1/3 of the wells containing a single cell (or more).

7. Alternatively, cells can be infected with enhanced green fluorescent protein (EGFP) expressing lentivirus and then collect EGFP expressing ECFC using a flow cytometry. Culture and expand the cells as described in the section under cell culture and expansion.

8. Tilt the plate at a 45-degree angle, touch the edge of the well with the multichannel pipette tip and gently aspirate.

9. Dilute DAPI stock solution 1:100 in PBS, and then 1:10 in deionized water.

10. Other nuclear staining dye such as SYTOX, Hoechst 33342, or Hoechst 33528 at 1 μg/ml concentration or can be used.

11. Cells in PBS can be stored at 4 °C for a maximum of 2 weeks.

References

1. Carmeliet P (2003) Angiogenesis in health and disease. Nat Med 9(6):653–660. https://doi.org/10.1038/nm0603-653

2. Eelen G, de Zeeuw P, Treps L, Harjes U, Wong BW, Carmeliet P (2018) Endothelial cell metabolism. Physiol Rev 98(1):3–58. https://doi.org/10.1152/physrev.00001.2017

3. Marcu R, Choi YJ, Xue J, Fortin CL, Wang Y, Nagao RJ, Xu J, MacDonald JW, Bammler TK, Murry CE, Muczynski K, Stevens KR, Himmelfarb J, Schwartz SM, Zheng Y (2018) Human organ-specific endothelial cell heterogeneity. iScience 4:20–35. https://doi.org/10.1016/j.isci.2018.05.003

4. Nolan DJ, Ginsberg M, Israely E, Palikuqi B, Poulos MG, James D, Ding BS, Schachterle W, Liu Y, Rosenwaks Z, Butler JM, Xiang J, Rafii A, Shido K, Rabbany SY, Elemento O, Rafii S (2013) Molecular signatures of tissue-specific microvascular endothelial cell heterogeneity in organ maintenance and regeneration. Dev Cell 26(2):204–219. https://doi.org/10.1016/j.devcel.2013.06.017

5. Aird WC (2007) Phenotypic heterogeneity of the endothelium: I. Structure, function, and mechanisms. Circ Res 100(2):158–173. https://doi.org/10.1161/01.RES.0000255691.76142.4a

6. Aird WC (2007) Phenotypic heterogeneity of the endothelium: II. Representative vascular beds. Circ Res 100(2):174–190. https://doi.org/10.1161/01.RES.0000255690.03436.ae

7. Hong Y, Eleftheriou D, Klein NJ, Brogan PA (2015) Impaired function of endothelial progenitor cells in children with primary systemic vasculitis. Arthritis Res Ther 17:292. https://doi.org/10.1186/s13075-015-0810-3

8. Lukowski SW, Patel J, Andersen SB, Sim SL, Wong HY, Tay J, Winkler I, Powell JE, Khosrotehrani K (2019) Single-cell transcriptional profiling of aortic endothelium identifies a hierarchy from endovascular progenitors to differentiated cells. Cell Rep 27(9):2748-2758. e2743. https://doi.org/10.1016/j.celrep.2019.04.102

9. Paik DT, Tian L, Lee J, Sayed N, Chen IY, Rhee S, Rhee JW, Kim Y, Wirka RC, Buikema JW, Wu SM, Red-Horse K, Quertermous T, Wu JC (2018) Large-scale single-cell RNA-seq reveals molecular signatures of heterogeneous populations of human induced pluripotent stem cell-derived endothelial cells. Circ Res 123(4):443–450. https://doi.org/10.1161/circresaha.118.312913

10. Veerman K, Tardiveau C, Martins F, Coudert J, Girard JP (2019) Single-cell analysis reveals heterogeneity of high endothelial venules and different regulation of genes controlling lymphocyte entry to lymph nodes. Cell Rep 26 (11):3116-3131.e3115. https://doi.org/10.1016/j.celrep.2019.02.042

11. Wakabayashi T, Naito H, Suehiro JI, Lin Y, Kawaji H, Iba T, Kouno T, Ishikawa-Kato S, Furuno M, Takara K, Muramatsu F, Weizhen J, Kidoya H, Ishihara K, Hayashizaki Y, Nishida K, Yoder MC, Takakura N (2018) CD157 marks tissue-resident endothelial stem cells with homeostatic and regenerative properties. Cell Stem Cell 22(3):384–397. https://doi.org/10.1016/j.stem.2018.01.010

12. Yoder MC (2018) Endothelial stem and progenitor cells (stem cells): (2017 Grover Conference Series). Pulm Circ 8 (1):2045893217743950. https://doi.org/10.1177/2045893217743950

13. Schniedermann J, Rennecke M, Buttler K, Richter G, Stadtler AM, Norgall S, Badar M, Barleon B, May T, Wilting J, Weich HA (2010) Mouse lung contains endothelial progenitors with high capacity to form blood and lymphatic vessels. BMC Cell Biol 11:50. https://doi.org/10.1186/1471-2121-11-50

14. Fang S, Wei J, Pentinmikko N, Leinonen H, Salven P (2012) Generation of functional blood vessels from a single c-kit+ adult vascular endothelial stem cell. PLoS Biol 10(10): e1001407. https://doi.org/10.1371/journal.pbio.1001407

15. Naito H, Kidoya H, Sakimoto S, Wakabayashi T, Takakura N (2012) Identification and characterization of a resident vascular stem/progenitor cell population in preexisting blood vessels. EMBO J 31(4):842–855. https://doi.org/10.1038/emboj.2011.465

16. Yu QC, Song W, Wang D, Zeng YA (2016) Identification of blood vascular endothelial stem cells by the expression of protein C receptor. Cell Res 26(10):1079–1098. https://doi.org/10.1038/cr.2016.85

17. Medina RJ, Barber CL, Sabatier F, Dignat-George F, Melero-Martin JM, Khosrotehrani K, Ohneda O, Randi AM, Chan JKY, Yamaguchi T, Van Hinsbergh VWM, Yoder MC, Stitt AW (2017) Endothelial progenitors: a consensus statement on nomenclature. Stem Cells Transl Med 6 (5):1316–1320. https://doi.org/10.1002/sctm.16-0360

18. Nowak-Sliwinska P, Alitalo K, Allen E, Anisimov A, Aplin AC, Auerbach R, Augustin HG, Bates DO, van Beijnum JR, Bender RHF, Bergers G, Bikfalvi A, Bischoff J, Bock BC, Brooks PC, Bussolino F, Cakir B, Carmeliet P, Castranova D, Cimpean AM, Cleaver O, Coukos G, Davis GE, De Palma M, Dimberg A, Dings RPM, Djonov V, Dudley AC, Dufton NP, Fendt SM, Ferrara N, Fruttiger M, Fukumura D, Ghesquiere B, Gong Y, Griffin RJ, Harris AL, Hughes CCW, Hultgren NW, Iruela-Arispe ML, Irving M, Jain RK, Kalluri R, Kalucka J, Kerbel RS, Kitajewski J, Klaassen I, Kleinmann HK, Koolwijk P, Kuczynski E, Kwak BR, Marien K, Melero-Martin JM, Munn LL, Nicosia RF, Noel A, Nurro J, Olsson AK, Petrova TV, Pietras K, Pili R, Pollard JW, Post MJ, Quax PHA, Rabinovich GA, Raica M, Randi AM, Ribatti D, Ruegg C, Schlingemann RO, Schulte-Merker S, Smith LEH, Song JW, Stacker SA, Stalin J, Stratman AN, Van de Velde M, van Hinsbergh VWM, Vermeulen PB, Waltenberger J, Weinstein BM, Xin H, Yetkin-Arik B, Yla-Herttuala S, Yoder MC, Griffioen AW (2018) Consensus guidelines for the use and interpretation of angiogenesis assays. Angiogenesis. https://doi.org/10.1007/s10456-018-9613-x

19. Ingram DA, Mead LE, Tanaka H, Meade V, Fenoglio A, Mortell K, Pollok K, Ferkowicz MJ, Gilley D, Yoder MC (2004) Identification of a novel hierarchy of endothelial progenitor cells using human peripheral and umbilical cord blood. Blood 104(9):2752–2760. https://doi.org/10.1182/blood-2004-04-1396

20. Ingram DA, Mead LE, Moore DB, Woodard W, Fenoglio A, Yoder MC (2005) Vessel wall-derived endothelial cells rapidly proliferate because they contain a complete hierarchy of endothelial progenitor cells. Blood 105(7):2783–2786. https://doi.org/10.1182/blood-2004-08-3057

21. Munoz-Hernandez R, Miranda ML, Stiefel P, Lin RZ, Praena-Fernandez JM, Dominguez-Simeon MJ, Villar J, Moreno-Luna R, Melero-Martin JM (2014) Decreased level of cord blood circulating endothelial colony-forming cells in preeclampsia. Hypertension 64(1):165–171. https://doi.org/10.1161/hypertensionaha.113.03058

22. Go E, Tarnawsky SP, Shelley WC, Banno K, Lin Y, Gil CH, Blue EK, Haneline LS, O'Neil KM, Yoder MC (2018) Mycophenolic acid induces senescence of vascular precursor cells. PLoS One 13(3):e0193749. https://doi.org/10.1371/journal.pone.0193749

23. Ingram DA, Krier TR, Mead LE, McGuire C, Prater DN, Bhavsar J, Saadatzadeh MR, Bijangi-Vishehsaraei K, Li F, Yoder MC, Haneline LS (2007) Clonogenic endothelial progenitor cells are sensitive to oxidative stress. Stem Cells 25(2):297–304. https://doi.org/10.1634/stemcells.2006-0340

24. Fraineau S, Palii CG, McNeill B, Ritso M, Shelley WC, Prasain N, Chu A, Vion E, Rieck K, Nilufar S, Perkins TJ, Rudnicki MA, Allan DS, Yoder MC, Suuronen EJ, Brand M (2017) Epigenetic activation of pro-angiogenic signaling pathways in human endothelial progenitors increases vasculogenesis. Stem Cell Rep 9(5):1573–1587. https://doi.org/10.1016/j.stemcr.2017.09.009

25. Basile DP, Yoder MC (2014) Circulating and tissue resident endothelial progenitor cells. J Cell Physiol 229(1):10–16. https://doi.org/10.1002/jcp.24423

26. Paschalaki KE, Randi AM (2018) Recent advances in endothelial colony forming cells toward their use in clinical translation. Front Med 5:295. https://doi.org/10.3389/fmed.2018.00295

27. O'Neill CL, McLoughlin KJ, Chambers SEJ, Guduric-Fuchs J, Stitt AW, Medina RJ (2018) The vasoreparative potential of endothelial colony forming cells: a journey through pre-clinical studies. Front Med 5:273. https://doi.org/10.3389/fmed.2018.00273

28. Smadja DM, Melero-Martin JM, Eikenboom J, Bowman M, Sabatier F, Randi AM (2019) Standardization of methods to quantify and culture endothelial colony-forming cells derived from peripheral blood: Position paper from the International Society on Thrombosis and Haemostasis SSC. J Thromb Haemost 17(7):1190–1194. https://doi.org/10.1111/jth.14462

29. Yoder MC, Mead LE, Prater D, Krier TR, Mroueh KN, Li F, Krasich R, Temm CJ, Prchal JT, Ingram DA (2007) Redefining endothelial progenitor cells via clonal analysis and hematopoietic stem/progenitor cell principals. Blood 109(5):1801–1809. https://doi.org/10.1182/blood-2006-08-043471

30. Wang CH, Hsieh IC, Su Pang JH, Cherng WJ, Lin SJ, Tung TH, Mei HF (2011) Factors associated with purity, biological function, and activation potential of endothelial colony-forming cells. Am J Physiol Regul Integr Comp Physiol 300(3):R586–R594. https://doi.org/10.1152/ajpregu.00450.2010

31. Critser PJ, Voytik-Harbin SL, Yoder MC (2011) Isolating and defining cells to engineer human blood vessels. Cell Prolif 44(Suppl 1):15–21. https://doi.org/10.1111/j.1365-2184.2010.00719.x

32. Colombo E, Calcaterra F, Cappelletti M, Mavilio D, Della Bella S (2013) Comparison of fibronectin and collagen in supporting the isolation and expansion of endothelial progenitor cells from human adult peripheral blood. PLoS One 8(6):e66734. https://doi.org/10.1371/journal.pone.0066734

33. Lin RZ, Hatch A, Antontsev VG, Murthy SK, Melero-Martin JM (2015) Microfluidic capture of endothelial colony-forming cells from human adult peripheral blood: phenotypic and functional validation in vivo. Tissue Eng Part C Methods 21(3):274–283. https://doi.org/10.1089/ten.TEC.2014.0323

## Chapter 4

In Vitro Coculture Assays of Angiogenesis

Haoche Wei, Ananthalakshmy Sundararaman, Sarah Line Bring Truelsen, David Gurevich, Jacob Thastrup, and Harry Mellor

Abstract

During angiogenesis, endothelial cells must undergo a coordinated set of morphological changes in order to form a new vessel. There is a need for endothelial cells to communicate with each other in order to take up different identities in the sprout and to migrate collectively as a connected chord. Endothelial cells must also interact with a wide range of other cells that contribute to vessel formation. In ischemic disease, hypoxic cells in tissue will generate proangiogenic signals that promote and guide angiogenesis. In solid tumors, this function is co-opted by tumor cells, which make a complex range of interactions with endothelial cells, even integrating into the walls of vessels. In vessel repair, cells from the immune system contribute to the promotion and remodeling of new vessels. The coculture angiogenesis assay is a long-term in vitro protocol that uses fibroblasts to secrete and condition an artificial stromal matrix for tubules to grow through. We show here how the assay can be easily adapted to include additional cell types, facilitating the study of cellular interactions during neovascularization.

Key words Endothelial cells, Angiogenesis, Coculture, Imaging, Tumor angiogenesis, Macrophages

1 Introduction

The coculture assay of angiogenesis was first described by Bishop et al. [1]. It was designed to mimic the three-dimensional processes of angiogenesis in vitro and to be used as a screen for inhibitors and activators of angiogenesis. In the original formulation of the assay, primary human umbilical endothelial cells (HUVEC) are mixed 50:50 with normal human dermal fibroblasts (NHDF) and plated onto fibronectin-coated dishes. Over 14 days, the HUVEC form a network of branching tubules that resemble capillaries [2]. The fibroblasts in the assay secrete a three-dimensional matrix that it is rich in collagen I and also contains fibronectin, tenascin-C, decorin, and versican [2, 3]. This mimics stromal matrix and triggers the endothelial cells into an angiogenic program. The mature assay is 3–5 cells deep, with endothelial tubes growing between layers of fibroblasts and matrix. This is deep enough for tube formation, but

Domenico Ribatti (ed.), *Vascular Morphogenesis: Methods and Protocols*, Methods in Molecular Biology, vol. 2206, https://doi.org/10.1007/978-1-0716-0916-3_4, © Springer Science+Business Media, LLC, part of Springer Nature 2021

shallow enough to enable high-resolution imaging. Importantly, the fibroblasts remodel the secreted matrix, producing fibrillar collagen. This gives an advantage over assays using simple gels of collagen or Matrigel. As the tubules mature, the endothelial cells secrete a basement lamina that is rich in laminin and collagen IV [2, 3]. Studies using electron microscopy show that the endothelial cells form a patent lumen [2].

Mavria et al. made an important modification to the original assay by plating the endothelial cells directly onto a confluent layer of fibroblasts. This shortens the time spent by the endothelial cells in the assay significantly, allowing for the use of oligonucleotide-based siRNA silencing [4]. This greatly facilitates screening for novel regulators of angiogenesis, with a high-content image-based readout [5, 6]. We have recently provided a detailed protocol for this, including the use of RNAi techniques for targeted gene suppression [7]. During angiogenesis, endothelial cells make interactions with a range of supporting cells types. Mural cells like pericytes become associated with maturing tubules to help stabilize the new vessel. Neuronal cells can provide guidance signals to endothelial cells to direct the direction of vessel growth. The coculture assay is amenable to the introduction of other cell populations, allowing the in vitro imaging of the interactions formed during vessel formation. Recently, we have used this assay to examine the contributions of different classes of macrophages during vessel repair [8]. We have also adapted the assay to examine interactions with patient-derived tumor organoids (tumoroids; unpublished data). Here we give methods for the introduction of macrophages and tumor cells into the coculture assay. The adaptations illustrate the flexibility of the assay, and its value as a bridge between in vitro and in vivo studies.

2   Materials

2.1   Cell Culture

1. Prepare HUVEC from human umbilical cords by standard methods [8] or purchase from Lonza.

2. Normal human dermal fibroblasts (Lonza).

3. EGM-2 medium (Lonza).

4. Opti-MEM medium (Life Technologies).

5. Ham's F12/DMEM and DMEM (Sigma).

6. Human fibronectin (Sigma).

2.2   Reagents for Cell Staining

1. PECAM-1 (CD31) primary antibody (R&D Systems #BBA7).

2. MCSFR primary antibody (Abcam # ab183316).

3. Alkaline phosphatase–conjugated secondary antibody (Novus Biologicals #NB720-AP).

4. Alexa Fluor 488 goat anti-mouse secondary antibody (Invitrogen #A11001).

5. SigmaFAST BCIP/NBT ((5-bromo-4-chloro-3-indolyl phosphate/nitro blue tetrazolium; Sigma #B5655).

6. 4% paraformaldehyde (PFA) in PBS, made fresh by heating mixture in a fume hood with stirring until dissolved and then letting cool to room temperature.

7. 10% Neutral buffered formalin with 4% formaldehyde.

8. Triton X-100.

9. DAPI (4′,6-diamidino-2-phenylindole).

3   Methods

3.1   Cell Culture

1. Maintain HUVEC in EGM-2 medium (Lonza) on plates coated with human fibronectin (10 µg/ml for 30 min). HUVEC are used up to passage 5.

2. Maintain NHDF in DMEM with 10% fetal calf serum (FBS), 100 U/ml penicillin, 100 µg/ml streptomycin and 292 µg/ml L-glutamine. NHDF are used up to passage 12.

3.2   The Standard Coculture Assay

1. Day 1: harvest NHDF by trypsinization and dilute in DMEM +10% FBS.

2. Count the cell density in a hemocytometer and adjust to 3×10^4 cells/ml in EGM-2.

3. Seed the cells into either 12-well tissue culture dishes or onto glass coverslips, as appropriate. There is no need to coat the surface. Tumoroid assays can be carried out in 24-well dishes.

4. Refresh the medium with EGM-2 on Day 4.

5. Day 5: the NHDF should now be confluent. Harvest the HUVEC by trypsinization and dilute in Ham's F12/DMEM +20% FBS.

6. Collect the HUVEC by gentle centrifugation at $700 \times g$ for 5 min. Resuspend them in EGM-2 and adjust the concentration to 3×10^4 cells/ml.

7. Seed onto the fibroblasts at a final concentration of 3×10^4 cells per well of a 12-well plate.

8. Refresh the medium with EGM-2 on Day 7 and Day 9.

9. The assay is complete on Day 11.

3.3   Staining Cocultures for Quantification

1. Aspirate the medium and fix the coculture using 70% ethanol at $-20\ °C$ for 30 min.

2. Incubate with 0.3% hydrogen peroxide in methanol for 15 min to remove endogenous alkaline phosphatase activity.

3. Wash three times with PBS and then incubate with mouse anti-CD31 antibody (0.25 μg/ml) in 1% BSA for 1 h at 37 °C.

4. Wash three times with PBS and then incubate with 0.6 μg/ml alkaline phosphatase–conjugated secondary antibody in 1% BSA for 1 h at 37 °C.

5. Wash six times in water and then add BCIP/NBT substrate (1 tablet in 10 ml water, 0.2 μm filtered).

6. Allow the stain to develop for 15–30 min at 37 °C and then wash the cells four times with water (prior to air-drying).

7. Image at low magnification without phase contrast (*see* **Notes 1–4**).

3.4 Preparation of Cocultures for Immunofluorescence Microscopy

1. Wash the cocultures three times in PBS and fix in 4% PFA in PBS for 15 min.

2. Wash again and permeabilize in 0.2% Triton X-100 in PBS for 5 min.

3. Wash again and incubate with fresh 0.5% (w/v) sodium borohydride in PBS to reduce autofluorescence.

4. Wash three times in PBS and then incubate with primary antibodies for 1 h in 1% BSA/PBS.

5. Wash three times in PBS and then incubate in fluorescent secondary antibody for 1 h in PBS.

6. Wash three times in PBS. A short incubation with DAPI can be used here (5 min, 1 μg/ml in water) to reveal nuclei if required. Mount coverslips and image (*see* **Note 5**). For tumoroid assays, cells are overlaid with PBS and imaged in the wells.

3.5 Preparation of Human Macrophages and Incorporation into Assay

1. Macrophages are prepared from human peripheral blood apheresis cones. Blood from apheresis cones is diluted 1:1 with Hanks balanced salt solution (Sigma) containing 0.6% acid citrate dextrose (ACD) and layered onto Histopaque (Sigma, ρ1.077). Samples are centrifuged at $400 \times g$ for 30 min without braking to generate a density gradient.

2. Macrophages are isolated from the interphase layer of the density gradient. CD14+ cells are isolated using CD14 MicroBeads, LS columns and a MidiMACS™ separator (Miltenyi Biotec), according to the manufacturer's instructions. Cells can be frozen in 50% FBS, 40% PBS with 10% DMSO (Sigma) and stored in liquid nitrogen until required.

3. Thawed cells are resuspended at a density of 0.33×10^6/ml in RPMI 1640 (Gibco) supplemented with 10% FBS (Gibco), 10 ng/ml macrophage colony-stimulating factor (Miltenyi Biotec). Cells are cultured with or without the inclusion of 20 ng/ml interleukin-4 (BioLegend) or 2.5 ng/ml interferon-γ (BioLegend), for directed differentiation into pro- and anti-inflammatory macrophages, respectively [8].

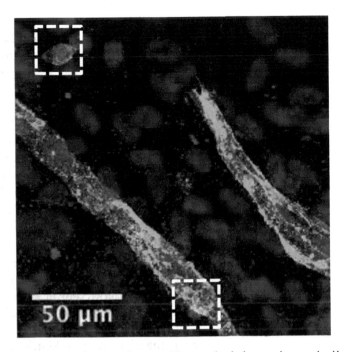

Fig. 1 Interactions of macrophages with vessels during angiogenesis. Human primary macrophages (red) were incubated with the coculture assay. The effects on vessel growth were determined by quantification of tubule density. Individual interactions between macrophages and vessels were observed using confocal microscopy

4. Cells are added to the coculture on Day 8 of the protocol. The medium is removed from the coculture and the macrophages added at a final concentration of $7.5 \times 10^4 \, \text{ml}^{-1}$. The assays are incubated for a further 3 days prior to fixation and staining. Macrophages are detected using mouse monoclonal anti-human MCSFR antibody at a dilution of 1:200 (Fig. 1).

3.6 Preparation of Tumoroids and Incorporation into Assay

1. Fresh tumor tissue is obtained from surgically resected tumors. Patient-derived tumoroids are prepared essentially as described by [9]. Briefly, freshly resected tumor samples of 0.01–1 g are cut into 1 mm^3 pieces and digested for 20 min at 37 °C with 1 mg/ml collagenase type II (Thermo Fisher Scientific, #17101-015) diluted in STEM medium (DMEM/F12, 1.8% BSA, 8 ng/ml bFGF (Thermo Fisher Scientific, #PHG0264), 200 U/ml penicillin, 200 μg/ml streptomycin, 100 μg/ml gentamicin, and StemPro® hESC diluted 1:50 (StemPro hESC SFM kit, Thermo Fisher Scientific, #A10008-01). The tissue suspension was passed through a 70 μm cell strainer and a 30 μm preseparation filter (Miltenyi Biotec). Tissue retained by the 70 μm filter was collected and redigested for 10 min at 37 °C and passed through the filters again. The tumor fragments (30–70 μm) were seeded in STEM cell medium in tissue

culture dishes coated with 1.5% agarose (Sigma-Aldrich) and cultured at 37 °C in a humidified incubator. After 3–5 days, tumoroids can be collected for further culture.

2. After 3–5 days, tumoroids sized 30–70 μm are seeded in drops of 50 μl Matrigel (VWR, #734-0271) diluted 1:2 with STEM cell medium at the bottom of a 24-well dish. After setting of the Matrigel for 30 min in a 37 °C incubator, the embedded tumoroids are overlaid with 1 ml STEM cell medium. After an average of 2 weeks, with intermediate medium change, the tumoroids exceeding 200 μm in size can be collected for experiments using a 200 μm pluriStrainer (PluriSelect, Leipzig, Germany, #SKU 43-50200). Alternatively, tumoroids can be mechanically dissociated by pipetting up and down until they pass through a 70 μm filter.

3. Tumoroids (>200 μm) are introduced to the assay on Day 7. The tumoroids are collected by gentle pipetting with cutoff P-1000 tips followed by filtration through a 400 μm and then a 200 μm pluriStrainer filter. The 200–400 μm fraction of tumoroids is suspended in PBS and allowed to sediment without centrifugation. A weak Matrigel solution is prepared by diluting growth factor reduced Matrigel 1:4 in a DMEM/F12-based medium; R-F12 (DMEM/F12, 5% FBS, 0.45% BSA, 0.5% fatty acid free BSA, 0.1 mM 2-mercaptoethanol, 100 U/ml penicillin, 100 μg/ml streptomycin, 50 μg/ml gentamicin, and 2.5 μg/ml amphotericin B). You should use 250 μl for each well of a 24-well plate. This tumoroid suspension is then carefully pipetted over the top of the HUVEC-NHDF coculture and incubated at 37 °C for 30 min to ensure setting of the matrix solution (*see* **Note 5**). Subsequently, 250 μl R-F12 (for a 24-well plate) is gently added over the solidified matrix. This cell culture medium is refreshed on Day 9.

4. Cells are fixed in 4% formalin for 2.5 h at 4 °C, with one change of formalin after 30 min. Most of the Matrigel should liquify at 4 °C, allowing the matrix layer to be washed away (*see* **Note 6**). The cancer coculture assays is then stained and imaged as usual (Fig. 2).

4 Notes

1. Endothelial cells in the coculture assay should form a network of capillary-like structures. If the culture contains islands of cells, this is usually a sign that the fibroblasts were too old or were not plated at a high enough density to ensure a confluent monolayer before plating the endothelial cells.

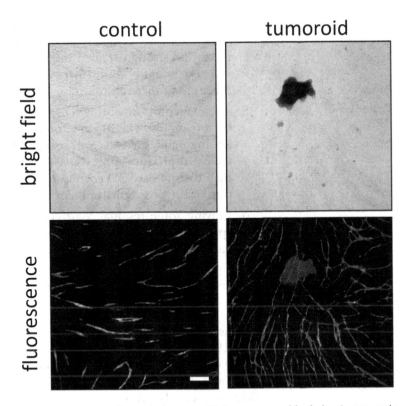

Fig. 2 Interactions of vessels (green) with tumor organoids during tumor angiogenesis. Human colon cancer organoids (red) were introduced into the coculture assay. The effect on vessel growth can be determined by quantification of tubule density. Individual interactions between vessels and tumoroids can be observed using confocal microscopy

2. Vessel growth can be quantified by measuring the length of vessels per unit area. This is better than attempting to quantify the length of individual vessels, which becomes impossible in denser cultures. A hemocytometer grid can be used to determine the area of the field of view. We use ImageJ to trace and analyze vessel length. A normal density should be 3–6 mm/mm^2.

3. Additional measurements include the number of branches formed per unit length. Branches occur both by splitting of an individual vessel, and by connection of two proximal vessels (anastomosis). As cultures become denser, the chances of two vessels meeting and forming a branch increases, and so it is important to remember that changes in the average length will affect branch number. Other parameters that can be quantified include vessel thickness and the number, length, and position of filopodia.

4. The culture is deep enough that antibody staining may be weaker for cells that are further from the surface. If you see a mixture of strongly stained and weakly stained cells, you should increase the antibody incubation times to ensure equal labelling throughout the culture.

5. The main purpose of the 3D Matrigel overlay is to immobilize the organoids. The 3D environment also supports their morphology and survival. Surprisingly, this version of the assay seems to produce more vessels and more extensive networks than the standard 2D coculture. This could reflect the generation of a hypoxic environment under the gel.

6. Leaving some residual Matrigel is not a problem, but it can appear on the images of the cocultures. The longer fixation is required due to the added thickness of the Matrigel layer.

Acknowledgments

This work was supported by project grants PG/16/62/32295 and PG/19/29/34319 from the British Heart Foundation and also by the Danish Innovation Fund, C.C. Klestrup Memorial Scholarship, Carl and Ellen Hertz Grant to Danish Medical and Natural Sciences, The Drost Fundation, Familien Hede Nielsens Fund, and Agnes and Poul Friis Fund.

References

1. Bishop ET, Bell GT, Bloor S, Broom IJ, Hendry NF, Wheatley DN (1999) An in vitro model of angiogenesis: basic features. Angiogenesis 3:335–344

2. Donovan D, Brown NJ, Bishop ET, Lewis CE (2001) Comparison of three in vitro human 'angiogenesis' assays with capillaries formed in vivo. Angiogenesis 4:113–121

3. Sorrell JM, Baber MA, Caplan AI (2007) A self-assembled fibroblast-endothelial cell coculture system that supports in vitro vasculogenesis by both human umbilical vein endothelial cells and human dermal microvascular endothelial cells. Cells Tissues Organs 186:157–168

4. Mavria G, Vercoulen Y, Yeo M, Paterson H, Karasarides M, Marais R, Bird D, Marshall CJ (2006) ERK-MAPK signaling opposes Rho-kinase to promote endothelial cell survival and sprouting during angiogenesis. Cancer Cell 9:33–44

5. Richards M, Hetheridge C, Mellor H (2015) The formin FMNL3 controls early apical specification in endothelial cells by regulating the polarized trafficking of podocalyxin. Curr Biol 25:2325–2331

6. Abraham S, Scarcia M, Bagshaw RD, McMahon K, Grant G, Harvey T, Yeo M, Esteves FOG, Thygesen HH, Jones PF, Speirs V, Hanby AM, Selby PJ, Lorger M, Dear TN, Pawson T, Marshall CJ, Mavria G (2015) A Rac/Cdc42 exchange factor complex promotes formation of lateral filopodia and blood vessel lumen morphogenesis. Nat Commun 6:7286

7. Richards M, Mellor H (2016) In vitro coculture assays of angiogenesis. Methods Mol Biol 1430:159–166

8. Gurevich DB, Severn CE, Twomey C, Greenhough A, Cash J, Toye AM, Mellor H, Martin P (2018) Live imaging of wound angiogenesis reveals macrophage orchestrated vessel sprouting and regression. EMBO J 37(13):pii: e97786

9. Jeppesen M, Hagel G, Glenthoj A, Vainer B, Ibsen P, Harling H, Thastrup O, Jørgensen LN, Thastrup J (2017) Short-term spheroid culture of primary colorectal cancer cells as an in vitro model for personalizing cancer medicine. PLoS One 12:e0183074

Chapter 5

Endothelial Cells: Co-culture Spheroids

Janos M. Kanczler, Julia A. Wells, and Richard O. C. Oreffo

Abstract

The development and maintenance of a functioning vascular system is a critical function for many aspects of tissue growth and regeneration. Vascular endothelial cell in vitro co-culture spheroids are self-organized cell composites that have the capacity to recapitulate the three-dimensional tissue microenvironment. These spheroid testing platforms aim to better understand the mechanisms of functional tissue and how new therapeutic agents can drive these 3D co-culture processes. Here we describe direct cell–cell 3D endothelial co-culture spheroid methods, to examine the physiological spatial growth and cell–cell interaction of vascular cells and surrounding native tissue cells in the formation of vascular networks within spheroids and the potential to regenerate tissue.

Key words Endothelial cells, Co-culture, Human bone marrow stromal cells, Spheroids, 3D, Tissue engineering

1 Introduction

Endothelial cells play a significant role in tissue growth and repair. Endothelial cells are pivotal in the coordination and physiological response to the constant demands and changes present within the circulatory system and associated vascular cell populations. Ideally suited to this vascular function as the primary component of the blood vessels, the endothelial cell plays a central role in the development and repair of functioning tissues and organs. Endothelial cells, classically interact with vascular smooth muscle cells to modulate vascular tone through signaling factors as well as through direct contact stimuli [1]. To facilitate vascularization, endothelial cells form tubelike structures during the progression of angiogenesis, initiating the process of new tissue vascularization. Nonetheless, a functional blood vessel system that can conduct blood flow cannot be readily created by the endothelial cells alone. Cells such as pericytes and vascular smooth muscle cells are required to functionalize the vasculature [2]. This complex relationship between

Domenico Ribatti (ed.), *Vascular Morphogenesis: Methods and Protocols*, Methods in Molecular Biology, vol. 2206, https://doi.org/10.1007/978-1-0716-0916-3_5, © Springer Science+Business Media, LLC, part of Springer Nature 2021

endothelial cells and the surrounding different tissue cell types, is pivotal in modulating and repairing the extracellular matrices [3, 4].

2D co-cultures using direct contact or non-contact transwells has provided valuable insights into the modulation of cell differentiation. In order to advance these co-culture models further, cell biologists and tissue engineers have recognized that three-dimensional (3D) spheroid models are an important tool in the study of cell–cell interactions with the extracellular matrix [5–7]. The complex interaction of the cells and extracellular matrix and, the close proximity of signaling proteins within a 3D environment mimic the cellular tissue growth. Over two decades ago, Korff and colleagues developed an endothelial cell (EC) spheroid 3D in vitro model to assess angiogenic responses and sprouting behavior [8]. Moreover, co-culture spheroids were generated using the "hanging drop" method or, round-bottom well plates, to replicate blood vessel physiology in order to better understand the close interactive mechanisms associated with endothelial cells and vascular smooth muscle cells [9].

These co-culture spheroid techniques have been replicated in other tissue types to further understand the close interaction of endothelial cells and the inherent local tissue cell populations [10]. An example of the 3D interaction of endothelial cells with its surrounding tissue specific cells is the development of endochondral ossification. Endochondral ossification necessitates the proliferation and differentiation of osteogenic and vascular cell precursors and the coordinated and temporal interactions of bone and cartilage formation together with the development of a functional vasculature critical in the developmental mechanisms and repair of endochondral bones [11, 12]. Thus, 3D spheroid endothelial co-culture models offer an attractive model to investigate the complex cell interaction and patterning of endothelial populations and musculoskeletal cells to induce functional osteoid matrix and bone formation during the process of endochondral bone formation [13]. This is exemplified in the chondrogenic cell priming induced prior to 3D endothelial cell/HBMSC spheroid generation, observed to modulate and significantly enhance osteogenic activity [14].

The use of co-cultured spheroids consisting of endothelial and mesenchymal cell progenitors can be assessed for in vitro angiogenesis [15]. Conversely, 3D organoids derived from human induced pluripotent stem cells (hiPSC) also provide valuable information on host tissue cell interaction and induction of angiogenesis [16]. Moreover, co-culture 3D spheroids of HUVEC/HBMSC implanted into an embryonic day 18 chick femur and cultured for

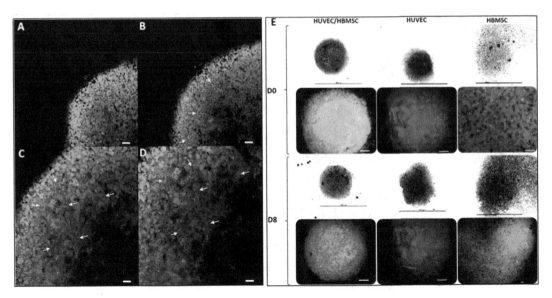

Fig. 1 (**a–d**) Confocal microscopy images of co-culture cell pellet constructs labeled with Vybrant™ Green: HBMSC, Red: HUVEC. Images show snapshots through the cell pellet constructs at day 8 of culture from top (**a**) to bottom (**d**) with HBMSCs dominating in the outer layer (**a**) of the pellet and with an increasing presence of HUVECs toward the center of the pellet (**b–d**); arrows depict areas of HUVEC aggregations; scale bar: 50 μm. Fluorescent and phase contrast images of cell constructs (labeled with Vybrant™) at D0 and D8 of in vitro organotypic culture (**e**). HBMSCs labeled with green fluorescence (Vybrant® CFDA); HUVECs labeled with red fluorescence (Vybrant™ DiI.). The co-culture pellet (HUVEC/HBMSC) display a more spherical, compact shape. HBMSC pellet, atypically, was not well formed and displayed more outgrowth of cells onto the surface. Scale bar fluorescent images: 100 μm; scale bar b/w images: 2000 μm (Fig. 1a–d reproduced from ref. 6 with permission from FASEB J) (Fig. 1e reproduced from ref. 19 with permission from University of Southampton)

10 days in an organotypic platform, demonstrated significant levels of bone repair as assessed via microcomputed X-ray tomography, suggesting a possible repair strategy for a critical-sized bone defect model [6] (Fig. 1).

Nevertheless, it is important to recognize, despite decades of research in the field of tissue regenerative medicine, the codevelopment of a vascular supply in a 3D tissue environment remains one of the key stumbling blocks in strategies for the repair and regeneration of tissue and organs at scale.

The experimental details and endothelial co-culture spheroid protocols are presented in the following sections. Examples of human bone marrow stromal cell (HBMSC)/endothelial (HUVEC) co-culture spheroids are detailed for the purposes of this chapter. However, other vascularized tissue cells may be used to produce co-culture spheroids to test endothelial cell culture interaction and functionality with other tissue-specific cells in a 3D spheroid environment.

2   Materials

First of all, the collection of human material as a source of endothelial cells must have signed consent from the person donating the tissue sample and requisite approval from national, local ethics and research governance (University) and National Research Ethics Service (NRES) must be in place (*see* **Note 1**).

2.1   Tissue

Human umbilical cords for isolation of endothelial cells (HUVECs) (*see* **Note 2**).

Human bone marrow (isolation of human bone marrow stromal cells [HBMSCs]) (*see* **Note 3**).

2.2   Reagents for Isolation and Culture of HUVECs

1× Phosphate Buffered Saline.

Collagenase Type B (5 mg/ml) diluted in sterile 1× PBS.

MEM 199.

Fetal calf serum.

Endothelial cell growth supplement + Heparin.

Sterile flow cabinet (Class II).

Sterile instruments, Forceps, N°22 surgical blade, Cannulas (approx. 2–3 mm diameter).

Sterile dish (for supporting umbilical cord).

Cable ties.

70% Ethanol (EtOH).

2.3   Reagents for HBMSC Isolation and Culture

70 μm cell strainer.

50 ml Falcon Tubes.

Fetal calf serum.

Alpha MEM (minimum essential medium α-modification).

2.4   Reagents for Monoculture and Co-culture Spheroids

Hemocytometer.

Centrifuge.

15 ml Eppendorf tubes (sterile).

2 ml microfuge tubes (sterile) (optional).

HUVEC culture medium: Endothelial cell Medium 199, 10% fetal calf serum, 0.4% (v/v) Endothelial cell growth Supplement/Heparin (ECGS/H); 1% of 10,000 IU/ml penicillin–streptomycin (P/S).

HBMSC culture medium: Alpha-MEM (minimum essential medium α-modification).

10% fetal calf serum, 1% of 10,000 IU/ml penicillin–streptomycin (P/S).

Vybrant® CFDA SE Cell Tracer Kit V12883 and Vybrant™ DiI cell labeling solution.

V22885.

3 Methods

3.1 Isolation and Culture of HUVEC

1. Human umbilical cords are typically collected in sterile containers after birth. The collection of samples for research are authorized under an ethically approved study with signed consent from the mothers. HUVECs are isolated and cultured as previously described by Jaffe 1973 with minor modifications [17].

2. The umbilical cord is cleaned using sterile swabs and 70% EtOH and examined for needle-stick injuries and clamp marks. A small section of tissue is removed at each end of the cord using a sterile scalpel.

3. Insert a cannula into the umbilical vein and secure using ×2 cable ties. Attach a 20 ml Luer Lock syringe and flush sterile 1× PBS slowly through the vein to remove any blood clots. Secure the other end of the cord (×2 cable ties) to create a sealed vein.

4. The umbilical cord is infused with sterile 5 mg/ml (w/v) of Collagenase B solution (Roche Diagnostics, Burgess Hill, UK) (*see* **Note 4**). Ensure that the vein is infused with collagenase such that the vein remains taut (*see* **Note 5**). Incubate the cord for 1 h (*see* **Notes 6** and **7**).

5. After the incubation period, withdraw the solution from the vein into the syringe and carefully fill the vein again repeating this **step 2** or **3** times. Remove the collagenase solution and place in a sterile 50 ml falcon tube.

6. An equal volume of endothelial cell culture medium supplemented with 1% P/S, 10% FCS, and ECGS/Heparin is added to the cell suspension and centrifuged at $110 \times g$ for 5 min.

7. Discard the supernatant and resuspend the pellet with fresh endothelial cell culture medium and plate into a filter-cap T-25 flask and culture at 37 °C, 5% CO_2 balanced air.

8. Grow and expand the cells using standard passaging with 1× trypsin–EDTA.

3.2 Isolation and Culture of HBMSC

1. Unselected HBMSC are isolated and cultured according to Oreffo et al. [18]. In a sterile class II cabinet, add 10mls of alpha MEM Tissue Culture Media (TCM) supplemented with 1× penicillin–streptomycin to the bone marrow (BM) aspirate

collected from patients undergoing elective total hip replacement. (NOTE the collection of samples for research must be authorized under an ethically approved study with signed consent from the patients and with full appropriate national ethical approval and under appropriate national guidelines.)

2. Vigorously shake the marrow and media solution and place through a 70 μm filter strainer to remove any fat and bone fragments. Repeat this process three times. Spin the marrow supernatant at $240 \times g$ for 5 min. Pour off the media and disperse the pellet of cells in 10 ml of 10% FCS in alpha MEM/$1\times$ P/S.

3. Transfer to a filter cap T-80 flask and culture at 37 °C 5% CO_2 balanced air.

4. Culture the cells and expand using standard passaging with $1\times$ trypsin–EDTA.

3.3 Generation of Co-culture 3D Spheroids

1. When sufficient quantities of the cell monolayers have been attained (*see* **Note 8**) for co-culture spheroid formation, harvest the cells using ($1\times$ trypsin–EDTA).

2. Undertake a cell count (hemocytometer) of the cells harvested to generate the seeding densities for the monoculture and co-culture spheroids. Centrifuge at $170 \times g$ for 4 min at 18 °C.

3. *Optional.* The cell types can be individually labeled to track them prior to the co-culture spheroids set up. HBMSCs are incubated in 10 mM CFDA (Vybrant® Cell Tracer Kit V12883) in 1 ml of $1\times$ PBS for 15 min at 37 °C. HBMSCs are repelleted and then resuspended in culture medium. The cell suspension is then incubated for 30 min at 37 °C, while the CFDA undergoes acetate hydrolysis. After incubation the cells are washed $2\times$ with $1\times$ PBS and transferred to 15 ml conical tubes. HUVECs are incubated in 5 μl/ml Vybrant™ Dil cell labeling solution (V22885) for 20 min at 37 °C, centrifuged at $170 \times g$ for 5 min and resuspended in culture medium. Cells are washed a further $2\times$ with $1\times$ PBS and a cell count performed (*see* **Note 9**).

4. For all cell cultures of the HUVEC with HBMSC, typically culture the pellets in 1:1 ratio of HUVEC TCM and HBMSC TCM (*see* **Notes 10** and **11**).

5. 2.5×10^5 cells each of HUVECs and HBMSC in 1 ml of either HUVEC: HBMSC (1:1) TCM for monoculture pellets or 1.25×10^5 cells each HUVECs/HBMSCs in 1 ml of co-culture HUVEC–HBMSC (1:1) TCM for co-culture pellets are transferred to 15 ml conical tubes (*see* **Note 12**). The cells are pelleted by centrifugation at $170 \times g$ for 5 min. Pellets are then incubated in the tube, leaving the lid slightly loose to

maintain oxygenation at 37 °C, 5% CO_2 for 2 days (*see* **Note 13**). Do not change the TCM during the 2 days of spheroid formation as this could disrupt the pellet.

6. These spheroids can be continuously grown within the tubes and applied to the research categories that the particular researchers are investigating. Culture media is changed using a sterile plastic Pastette (3 ml) by gently tilting the tube and withdrawing the excess TCM ensuring the spheroid is not disturbed. This is particularly important for the HUVEC monospheroids as these spheroids are fragile. Gently, on the side of the tilted tube, add fresh TCM slowly (*see* **Note 14**).

7. After a period of culture, the spheroids can be removed from the tube. The optimal way of removing the spheroid from the tube is by using a sterile Pastette. Firstly, place the Pastette just above the spheroid and slowly remove the TCM and pellet. The spheroids at this point are quite robust (*see* **Note 15**). Within the Pastette carefully allow the spheroid to migrate to the bottom of the tip. The spheroid can then be dispensed to another tube for fixation or molecular analysis. For example, after fixation with 4% PFA the localization of each cell type can be visualized using confocal fluorescent microscopy (Fig. 1) [6, 19].

8. Once the pellet has been generated, the pellet can be transferred to experimental studies. As illustrated in Fig. 2, HUVEC/HBMSC co-culture spheroids can be transferred to a bone defect and cultured by organotypic culture for 10 days.

9. The co-culture spheroids can be analyzed using a variety of techniques including; X-ray microcomputed tomography, confocal microscopy, qPCR, Western blotting, histology, and immunohistochemistry. Fixation can be through formalin or OCT depending on the final analytical techniques (*see* **Note 16**).

4   Notes

1. It is essential to obtain ethical approval from the guidelines established in the researcher's country of origin for the consent and use of the human cells/tissue for research.

2. Depending on the source of tissue, other endothelial cells can be used, including human aortic endothelial cells (HAoECs), endothelial progenitor cells (EPCs), human dermal microvascular endothelial cells (HDMECs), human coronary artery endothelial cells (HCAECs), and human endothelial progenitor cells (EPCs). In addition, the EA.hy926 (ATCC® CRL-2922™) cell line can be used in co-culture studies.

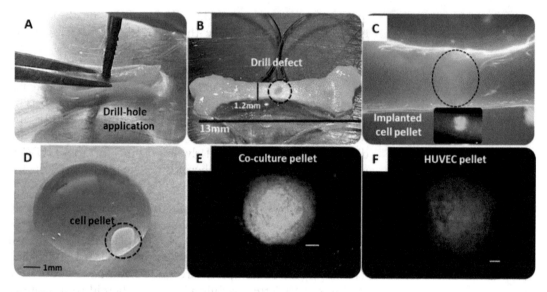

Fig. 2 Drill defect application and pellet implantation in E18 chick femurs. Representative images of the preparation of E18 chick femur drill defects implanted with cell pellets for a 10 days organ culture. Creation of a drill defect using a sterile 0.9 mm metal drill bit (**a**). Femur with applied drill defect (circle) (**b**). Femur with pellet implanted into the drill defect (circle) prior to the 10 days organotypic culture period (**c**). Cell pellet suspended in media in a petri dish prior to implantation (**d**). Sample image of fluorescently labeled co-culture cell spheres HBMSC/HUVEC (green, Vybrant carboxyfluorescein diacetate; red, Vybrant DiI) and HUVEC only (Vybrant DiI), at day 8 of in vitro culture (**e, f**). Scale bars, 100 µm (Fig. 2 reproduced from ref. 6 with permission from FASEB J)

3. We use human primary human bone marrow from patients undergoing elective orthopedic surgery. There are commercially available sources of HBMSC.

4. Ensure the collagenase B has been prewarmed to 37 °C. The volume infused into the vein is dependent on the size of the cord.

5. It is important that the collagenase solution is in contact with the endothelium surface.

6. If at all possible, place the prepared umbilical cord containing collagenase in a 37 °C incubator. If you do not have this facility the cord can be left in the Class II sterile cabinet at room temperature.

7. Continuously check the cord for any leakages particular around the cannula region.

8. Depending on the cell type used for the co-culture spheroids with endothelial cells ensure minimal passaging occurs to ensure negligible differentiation has been attained for primary cultured cells. In our studies using HBMSC, we typically do not use cells higher than passage 1. This is not a major issue if well characterized cell lines are used.

9. This step can be omitted if labeling of cells is not required for the experiment.

10. We have observed that in co-culture spheroids, cells respond better with a 1:1 mix of the two types of culture medium. For consistent experimental cultures, we have typically applied this 1:1 culture medium to the monocultured spheroids of HUVECs and HBMSCs as well.

11. There have been a number of different ratios used of co-culture endothelial cells with other cell types stated in the literature. Optimization may be required dependent on the cells that are being co-cultured with the endothelial cells to create 3D spheroids.

12. The number of cells and the ratio of the cells needs to be determined by the user. In addition, 3D spheroid formation can be undertaken in 1 ml sterile Eppendorf tubes (or even smaller tubes, e.g., 200 μl microtubes) This can be easier to transfer the final spheroid to other experimental studies.

13. There are other methods of forming spheroids including the "hanging drop" method [20] and the nonadherent cell culture plate model 96-w. BRAND plates—U inert Grade—clear—Ref: 781900.

14. These spheroids can be grown in organotypic culture systems where the spheroids are placed on a PTFE sterile membrane on 0.4 μm sterile filter with TCM just covering the filter paper so that the spheroid is cultured at an air–liquid interphase.

15. If the co-cultured spheroid has grown larger than the Pastette tip bore size, the tip can be cut further up the shaft to allow the spheroid to be drawn up into a larger nozzle region.

16. During the dehydration step for histology wax processing, the pellets can be small and relatively transparent, making it challenging to embed the sample. Prestaining with Eosin allows for the location of the sample in the wax block. Eosin does not interfere with histological or immunohistochemical stains when the spheroids are sectioned. This can also be applied to the process of cryo-sectioning.

Acknowledgements

Funding to ROCO from the Biotechnology and Biological Sciences Research Council UK (BBSRC LO21071/ and BB/L00609X/1), the UK Regenerative Medicine Platform Acellular/Smart Materials—3D Architecture (MR/R015651/1), and a grant from the UK Regenerative Medicine Platform (MR/L012626/1 Southampton Imaging) is gratefully acknowledged.

References

1. Lilly B (2014) We have contact: endothelial cell-smooth muscle cell interactions. Physiology 29(4):234–241

2. Sweeney M, Foldes G (2018) It takes two: endothelial-perivascular cell cross-talk in vascular development and disease. Front Cardiovasc Med 5:154

3. Lazzari G, Nicolas V, Matsusaki M, Akashi M, Couvreur P, Mura S (2018) Multicellular spheroid based on a triple co-culture: a novel 3D model to mimic pancreatic tumor complexity. Acta Biomater 78:296–307

4. Shoval H, Karsch-Bluman A, Brill-Karniely Y, Stern T, Zamir G, Hubert A, Benny O (2017) Tumor cells and their crosstalk with endothelial cells in 3D spheroids. Sci Rep 7(1):10428

5. Guerrero J, Oliveira H, Catros S, Siadous R, Derkaoui S-M, Bareille R, Letourneur D, Amédée J (2014) The use of total human bone marrow fraction in a direct three-dimensional expansion approach for bone tissue engineering applications: focus on angiogenesis and osteogenesis. Tissue Eng A 21(5–6):861–874

6. Inglis S, Kanczler JM, Oreffo RO (2018) 3D human bone marrow stromal and endothelial cell spheres promote bone healing in an osteogenic niche. FASEB J 33(3):3279–3290

7. Walser R, Metzger W, Görg A, Pohlemann T, Menger M, Laschke M (2013) Generation of co-culture spheroids as vascularisation units for bone tissue engineering. Eur Cell Mater 26:222–233

8. Korff T, Augustin HG (1998) Integration of endothelial cells in multicellular spheroids prevents apoptosis and induces differentiation. J Cell Biol 143(5):1341–1352

9. Pfisterer L, Korff T (2016) Spheroid-based in vitro angiogenesis model. In: Angiogenesis protocols, 3rd edn. Springer, New York, NY, pp 167–177

10. Chiew GGY, Wei N, Sultania S, Lim S, Luo KQ (2017) Bioengineered three-dimensional co-culture of cancer cells and endothelial cells: a model system for dual analysis of tumor growth and angiogenesis. Biotechnol Bioeng 114(8):1865–1877

11. Carano RA, Filvaroff EH (2003) Angiogenesis and bone repair. Drug Discov Today 8 (21):980–989

12. Kanczler J, Oreffo R (2008) Osteogenesis and angiogenesis: the potential for engineering bone. Eur Cell Mater 15(2):100–114

13. Bruder SP, Caplan AI (1989) Cellular and molecular events during embryonic bone development. Connect Tissue Res 20 (1–4):65–71

14. Freeman FE, Haugh MG, McNamara LM (2015) An in vitro bone tissue regeneration strategy combining chondrogenic and vascular priming enhances the mineralization potential of mesenchymal stem cells in vitro while also allowing for vessel formation. Tissue Eng A 21 (7–8):1320–1332

15. Shah S, Lee H, Park YH, Jeon E, Chung HK, Lee ES, Shim JH, Kang K-T (2019) Three-dimensional angiogenesis assay system using co-culture spheroids formed by endothelial colony forming cells and mesenchymal stem cells. J Vis Exp 151:e60032

16. Wimmer RA, Leopoldi A, Aichinger M, Kerjaschki D, Penninger JM (2019) Generation of blood vessel organoids from human pluripotent stem cells. Nat Protoc 14 (11):3082–3100

17. Jaffe EA, Nachman RL, Becker CG, Minick CR (1973) Culture of human endothelial cells derived from umbilical veins. Identification by morphologic and immunologic criteria. J Clin Invest 52(11):2745–2756

18. Oreffo RO, Driessens FC, Planell JA, Triffitt JT (1998) Growth and differentiation of human bone marrow osteoprogenitors on novel calcium phosphate cements. Biomaterials 19 (20):1845–1854

19. Inglis S (2017) The role of the vasculature in skeletal development and repair – clues to improved regenerative strategies. Ph.D. Southampton, Southampton

20. Hoarau-Véchot J, Rafii A, Touboul C, Pasquier J (2018) Halfway between 2D and animal models: are 3D cultures the ideal tool to study cancer-microenvironment interactions? Int J Mol Sci 19(1):181

Chapter 6

Microfluidic Device Setting by Coculturing Endothelial Cells and Mesenchymal Stem Cells

Masafumi Watanabe and Ryo Sudo

Abstract

The construction of vascular networks is essential for developing functional organ/tissue constructs in terms of oxygen and nutrient supply. Although recent advances in microfluidic techniques have allowed for the construction of microvascular networks using microfluidic devices, their structures cannot be maintained for extended periods of time due to a lack of perivascular cells. To construct long-lasting microvascular networks, it is important that perivascular cells are present to provide structural support to vessels, because in vivo microvessels are covered by perivascular cells and stabilized. Here, we describe a microfluidic cell culture platform for the construction of microvascular networks with supportive perivascular cells. Our results showed that microvascular networks covered by pericyte-like perivascular cells formed in a microfluidic device and their structures were maintained for at least 3 weeks in vitro.

Key words Endothelial cells, Mesenchymal stem cells, Microfluidic devices, Pericytes

1 Introduction

The formation of vascular networks is crucial for the maintenance of functional organs and tissues such as the liver, lungs, and kidneys because blood flow delivers oxygen and nutrients throughout organ/tissue constructs via vasculatures [1]. In the absence of vascular networks, large-scale tissue constructs, which are thicker than 100–200 μm, cannot maintain their viability in the long term because such tissue constructs may lead to hypoxia and cellular necrosis due to oxygen diffusion limitation [2–4]. Therefore, it is necessary to construct microvascular networks for the development of three-dimensional (3D) organ/tissue constructs in vitro.

In recent years, various 3D cell culture approaches, such as bioprinting, premold methods, and microfluidics, have been developed to construct vascular networks in vitro [5]. In particular, microfluidic techniques (e.g., microfluidic devices) are regarded as promising for vascular tissue engineering because microfluidic devices allow us to produce well-controlled culture environments

Domenico Ribatti (ed.), *Vascular Morphogenesis: Methods and Protocols*, Methods in Molecular Biology, vol. 2206, https://doi.org/10.1007/978-1-0716-0916-3_6, © Springer Science+Business Media, LLC, part of Springer Nature 2021

that mimic cellular microenvironments in vivo. Previous studies have demonstrated that endothelial cells (ECs) form vascular networks with continuous lumens in microfluidic devices [6–8]. However, these vascular networks are not stable and the maintenance of these structures is limited to a short culture period of 4–5 days. This might be due to the lack of perivascular cells, because in vivo vessels are supported by perivascular cells such as pericytes and smooth muscle cells [9]. Thus, formation of stable vascular networks with perivascular cells is needed to produce long-lasting organ/tissue constructs in vitro.

Here, we describe a microfluidic cell culture platform to construct microvascular networks with perivascular cells in the coculture of human umbilical vein endothelial cells (HUVECs) and mesenchymal stem cells (MSCs). Our results demonstrated that HUVECs extended microvascular networks covered by perivascular cells in HUVEC–MSC coculture [10]. In particular, MSCs differentiated into perivascular cells and expressed several pericyte markers such as α smooth muscle actin (αSMA), neural/glial antigen 2 (NG2), and platelet-derived growth factor β (PDGFRβ) in HUVEC–MSC coculture [11, 12]. Furthermore, vascular networks with continuous lumens with diameters less than 10 μm were formed and maintained for at least 3 weeks. This approach will provide new insights into the formation of long-lasting organ/tissue constructs in vitro in terms of vascularization.

2   Materials

2.1   Fabrication of Master Molds (Photolithography)

1. An epoxy-based photoresist designed for microfabrication, SU-8 3050 (Nippon Kayaku, Tokyo, Japan).

2. 4-inch silicon wafer.

3. Spin coater.

4. Hot plate.

5. Transparency photomask with a desired micropattern.

6. Mask aligner.

7. SU-8 developer.

8. Isopropyl alcohol (IPA).

9. Glass petri dish.

2.2   Fabrication of Microfluidic Devices (Soft Lithography)

1. SYLGARD® 184 silicone-elastomer base and the curing agent.

2. Glass rod.

3. Plastic cup.

4. Vacuum chamber.

5. Master mold with a desired micropattern.

6. Gas duster.

7. Oven.

8. Scalpel.

9. Razor blade.

10. Biopsy punch.

11. Scotch tape.

12. Glass beaker.

13. Deionized water.

14. Autoclavable container.

2.3 Formation of Microchannels and Surface Treatment

1. Autoclaved polydimethylsiloxane (PDMS) device.

2. Coverslip.

3. Plasma cleaner.

4. 60-mm dish

5. 1 mg/ml poly-D-lysine (PDL) in sterile deionized water (store at −20 °C).

6. Incubator.

7. Sterile deionized water.

8. Oven.

2.4 Gel Formation in the Gel Region

1. 10× PBS.

2. 0.5 N NaOH.

3. Sterile deionized water.

4. Rat tail collagen type I solution (*see* **Note 1**).

5. 10-μl micropipette.

6. 24-well plate (*see* **Note 2**).

7. Incubator.

2.5 HUVEC–MSC Coculture in a Microfluidic Device

1. HUVECs.

2. Trypsin–EDTA.

3. Endothelial cell growth medium-2 (EGM-2) (Lonza, Walkersville, MD).

4. MSCs (*see* ref. 13).

5. MSC medium: Dulbecco's modified Eagle's medium (DMEM) supplemented with 20% fetal bovine serum (FBS) and an antibiotic–antimycotic.

6. 100 μg/ml vascular endothelial growth factor (VEGF) in sterile deionized water (VEGF stock solution; store at −20 °C).

7. 100 µg/ml basic fibroblast growth factor (bFGF) in 5 mM Tris-HCl (bFGF stock solution; store at −20 °C).

8. HUVEC–MSC coculture medium: a 1:1 mixture of EGM-2 and MSC medium supplemented with 10 ng/ml bFGF and 10 ng/ml VEGF (*see* refs. 10, 11) (*see* **Note 3**).

3 Methods

3.1 Fabrication of Master Molds (Photolithography)

1. Dispense ~4 ml of SU-8 3050 onto a cleaned 4-inch silicon wafer set in a spin coater.

2. Coat SU-8 3050 uniformly on the wafer using a spin coater.

3. Prebake the coated wafer at 95 °C on a hot plate (*see* **Note 4**).

4. Expose the coated wafer to an appropriate dose of UV light through a transparency photomask using a mask aligner to produce a desired pattern on the silicon wafer (*see* **Note 5**).

5. After UV exposure, bake the wafer at 65 °C and then at 95 °C on the hot plate (*see* **Note 6**).

6. Develop in SU-8 developer and rinse the wafer (master mold) briefly with IPA (*see* ref. 14).

7. Place the master mold in a glass petri dish until use.

3.2 Fabrication of Microfluidic Devices (Soft Lithography)

1. Mix the SYLGARD® 184 silicone-elastomer base and the curing agent in a 10:1 ratio using a glass rod in a plastic cup.

2. Place the mixture, which is PDMS prepolymer, in a vacuum chamber for 30 min to remove air bubbles.

3. Pour the PDMS mixture onto the master mold up to 6–7 mm in height, and then, remove bubbles with a gas duster.

4. Place the master mold with the PDMS mixture in an oven at 65 °C for >6 h to cure PDMS (*see* **Note 7**).

5. Peel the cured PDMS from the master mold using a scalpel.

6. Cut the cured PDMS into individual devices using a razor blade and punch the PDMS device by a biopsy punch to form inlets and outlets of microchannels.

7. Remove small particles on the PDMS devices using Scotch tape.

8. Autoclave the PDMS devices in a glass beaker with deionized water.

9. Place the PDMS devices in an autoclavable clean container and autoclave them again.

10. Dry the autoclaved PDMS devices in the clean container in an oven at 65 °C.

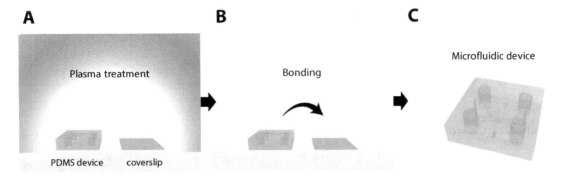

Fig. 1 Schematic diagrams of plasma bonding. (A) A PDMS device and a coverslip are exposed to air plasma using a plasma cleaner. (B) After plasma exposure, the PDMS device is bonded to the coverslip. (C) A microfluidic device with enclosed microchannels is fabricated

3.3 Formation of Microchannels and Surface Treatment

1. Place the autoclaved PDMS devices and sterile coverslips in a chamber of a plasma cleaner and then reduce pressure inside the chamber.

2. Expose the autoclaved PDMS devices and sterile coverslips to air plasma for 60 s (Fig. 1A) (*see* **Note 8**).

3. Bond a PDMS device to a coverslip for assembling a microfluidic device and gently press the PDMS device to enhance adhesion between the PDMS device and the coverslip (Fig. 1B, C).

4. Place the microfluidic device in a 60-mm dish.

5. Fill the gel region of the microfluidic device with 1 mg/ml PDL solution to enhance adhesion between PDMS (or coverslip) and gel (*see* **Note 9**).

6. Place the microfluidic devices in a humidified incubator (37 °C, 5% CO_2) for >6 h.

7. Aspirate the PDL solution and rinse the microchannels with sterile deionized water twice.

8. Aspirate the sterile deionized water completely in the microchannels and dry the microfluidic devices in an oven at 65 °C for >24 h.

3.4 Gel Formation in the Gel Region

1. Prepare collagen gel prepolymer at the desired concentration (*see* **Note 10**) by mixing 10× PBS, 0.5 N NaOH, sterile deionized water, and rat tail collagen type I solution (*see* **Note 11**). For example, 200 μl of 3 mg/ml collagen gel prepolymer can be made by mixing 20 μl 10× PBS, 5 μl 0.5 N NaOH, 55 μl sterile deionized water, and 120 μl of 5 mg/ml collagen solution (*see* **Note 12**).

2. Inject the collagen gel solution slowly from an inlet of the gel channel using a 10-μl micropipette (Fig. 2A, B).

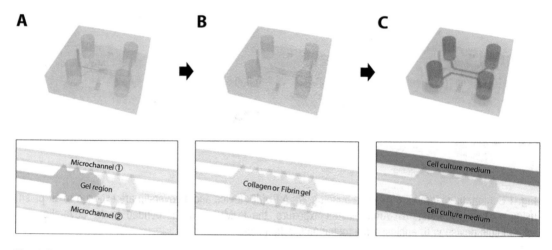

Fig. 2 Gel formation in the gel region. (A) Gel prepolymer (e.g., collagen, fibrin) is injected to fill the gel region through an inlet of the central channel. (B) The gel region is filled with the gel prepolymer. (C) After gelation, empty microchannels are filled with cell culture medium

3. Put the microfluidic devices on a 24-well plate filled with sterile deionized water in an inverted position to prevent evaporation of the gel.

4. Place the 24-well plate with the microfluidic devices in the incubator for 30 min for gelation.

5. Fill the medium channels gently with warmed cell culture medium and keep the microfluidic devices in the incubator until use (Fig. 2C).

3.5 HUVEC–MSC Coculture in a Microfluidic Device

1. Detach HUVECs from cell culture flasks using trypsin–EDTA and suspend them in EGM-2 at a concentration of 1×10^6 cells/ml.

2. Add 10 μl of the HUVEC suspension to an inlet of one of the medium channels (Fig. 3A, B).

3. Tilt the microfluidic devices vertically to allow for HUVEC attachment to the interface of the gel (Fig. 3C).

4. Place the microfluidic devices in the incubator for 30 min to promote cell attachment to the gel interface.

5. Detach MSCs from cell culture flasks using trypsin–EDTA and suspend them in MSC medium at a concentration of 1×10^6 cells/ml (*see* ref. [10, 11]).

6. Add 10 μl of the MSC suspension to an inlet of the medium channel (Fig. 3D).

7. Tilt the microfluidic devices vertically to allow for MSC attachment to the interface of the gel.

8. Place the microfluidic devices in the incubator for 30 min to promote cell attachment to the gel.

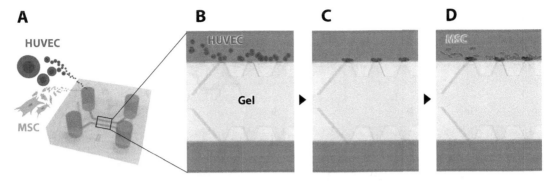

Fig. 3 Schematic diagrams of cell seeding in a microfluidic device. (A) HUVECs and MSCs are sequentially seeded into a microfluidic device. (B) The HUVEC suspension is added to a microchannel. (C) HUVECs attach to the interface of the gel. (D) MSC suspension is added to the microchannel in the same manner

9. Add coculture medium to inlets of two medium channels.

10. Change coculture medium daily in microfluidic devices.

11. Microvascular networks are formed in the gel region of microfluidic devices (Fig. 4A, B) (*see* **Note 13**).

12. Microvessels are covered by MSC-derived perivascular cells, which are positive for αSMA, after 1 week and their microstructures are maintained for at least 3 weeks (Fig. 5A, B) (*see* **Note 14**).

4   Notes

1. Rat tail type I collagen solution (7–9 mg/ml) is commercially obtained and used in experiments.

2. Pour sterile deionized water into a 24-well plate. Place this container in a 5% CO_2 incubator at 37 °C until use.

3. VEGF and bFGF stock solutions are added to HUVEC–MSC coculture medium to become final concentration.

4. Determine the heating time required for prebake according to the datasheet of SU-8 3050 (Microchem).

5. Determine UV exposure time required for cross-link according to the datasheet of SU-8 3050 (Microchem).

6. Determine the heating time required for postbake according to the datasheet of SU-8 3050 (Microchem).

7. The temperature of the oven can be increased to shorten the curing time of PDMS.

8. Exposure time of air plasma should be examined depending on the specification of the plasma cleaner.

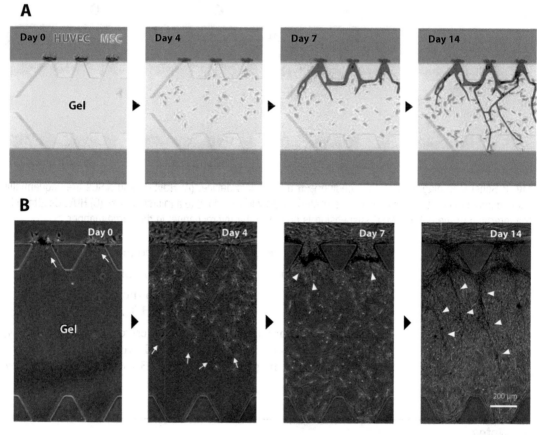

Fig. 4 Formation of microvascular networks in HUVEC–MSC coculture. (**a**) Schematic diagrams and (**b**) corresponding phase-contrast images in the process of microvascular network formation. HUVECs and MSCs attach to the interface of the gel on day 0 and then MSCs migrate into the gel on day 4 (arrows). HUVECs form vascular sprouts on day 7 and subsequently extend vascular networks on day 14 in the gel region (arrowheads). Scale bar = 200 μm

9. PDL coating of microchannels is carried out to avoid separation of collagen gel from channel walls. This process promotes 3D migration of cells in the collagen gel (*see* ref. 15).

10. All ingredients for collagen gel prepolymer should be kept on ice to avoid unexpected gelation.

11. Collagen type I solution at concentrations of 3–4 mg/ml was used in previous reports (*see* ref. 11, 16). In addition, fibrin gel at a concentration of 4 mg/ml also allowed for the formation of microvascular networks with pericyte-like cells (*see* ref. 12).

12. The amount of 0.5 N NaOH is adjusted to get pH 7.2–7.4 of the collagen gel prepolymer because the pH value during gelation affects the stiffness of the collagen gel, which may result in different cellular morphogenesis in the gel (*see* ref. 17).

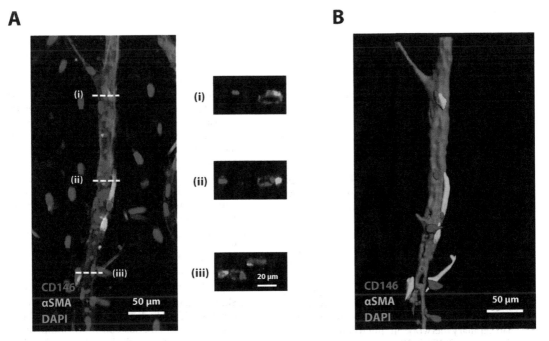

Fig. 5 Formation of microvessels covered by perivascular cells. (**a**) A projected confocal image shows that a microvessel (CD146, in red) is covered by perivascular cells (αSMA, in green) in HUVEC–MSC coculture. Cell nuclei are counter stained with 4′,6-diamidino-2-phenylindole (DAPI, in blue). Fluorescence images show continuous lumen of the microvessel. Cross-sectional images correspond to lines i–iii. (**b**) A 3D reconstructed image shows that microvessels are covered by αSMA-positive perivascular cells. Scale bars = 50 μm (**a, b**), = 20 μm (i–iii)

13. In HUVEC–MSC coculture, MSCs migrated into the gel in an early stage of vascular formation and HUVECs subsequently formed vascular sprouts and extended microvascular networks in the gel region.

14. We demonstrated that perivascular cells in HUVEC–MSC coculture were positive for αSMA, which is a pericyte marker (Fig. 5). Our previous studies further confirmed that these perivascular cells were also positive for the other pericyte markers such as NG2 and PDGFRβ (*see* refs. 11, 12).

Acknowledgments

This work was supported by Japan Society for Promotion of Science KAKENHI Grant Numbers 18K19937 and 19H04452 to R.S.

References

1. Sudo R (2014) Multiscale tissue engineering for liver reconstruction. Organogenesis 10:216–224

2. Carmeliet P, Jain RK (2000) Angiogenesis in cancer and other diseases. Nature 407:249–257

3. Jain RK, Au P, Tam J, Duda DG, Fukumura D (2005) Engineering vascularized tissue. Nat Biotechnol 23:821–823

4. Miller JS, Stevens KR, Yang MT, Baker BM, Nguyen DHT, Cohen DM et al (2012) Rapid casting of patterned vascular networks for perfusable engineered three-dimensional tissues. Nat Mater 11:768–774

5. Bogorad MI, DeStefano J, Karlsson J, Wong AD, Gerecht S, Searson PC (2015) Review: In vitro microvessel models. Lab Chip 15:4242–4255

6. Yeon JH, Ryu HR, Chung M, Hu QP, Jeon NL (2012) In vitro formation and characterization of a perfusable three-dimensional tubular capillary network in microfluidic devices. Lab Chip 12:2815–2822

7. Kim S, Lee H, Chung M, Jeon NL (2013) Engineering of functional, perfusable 3D microvascular networks on a chip. Lab Chip 13:1489–1500

8. Abe Y, Watanabe M, Chung S, Kamm RD, Tanishita K, Sudo R (2019) Balance of interstitial flow magnitude and vascular endothelial growth factor concentration modulates three-dimensional microvascular network formation. APL Bioeng 3:036102

9. Ribatti D, Nico B, Crivellato E (2011) The role of pericytes in angiogenesis. Int J Dev Biol 55:261–268

10. Yamamoto K, Tanimura K, Mabuchi Y, Matsuzaki Y, Chung S, Kamm RD et al (2013) The stabilization effect of mesenchymal stem cells on the formation of microvascular networks in a microfluidic device. J Biomech Sci Eng 8:114–128

11. Yamamoto K, Tanimura K, Watanabe M, Sano H, Uwamori H, Mabuchi Y et al (2019) Construction of continuous capillary networks stabilized by pericyte-like perivascular cells. Tissue Eng Part A 25:499–510

12. Uwamori H, Ono Y, Yamashita T, Arai K, Sudo R (2019) Comparison of organ-specific endothelial cells in terms of microvascular formation and endothelial barrier functions. Microvasc Res 122:60–70

13. Mabuchi Y, Morikawa S, Harada S, Niibe K, Suzuki S, Renault-Mihara F et al (2013) LNGFR+THY-1+VCAM-1hi+ cells reveal functionally distinct subpopulations in mesenchymal stem cells. Stem Cell Rep 1:152–165

14. Shin Y, Han S, Jeon JS, Yamamoto K, Zervantonakis IK, Sudo R et al (2012) Microfluidic assay for simultaneous culture of multiple cell types on surfaces or within hydrogels. Nat Protoc 7:1247–1259

15. Chung S, Sudo R, Zervantonakis IK, Rimchala T, Kamm RD (2009) Surface-treatment-induced three-dimensional capillary morphogenesis in a microfluidic platform. Adv Mater 21:4863–4867

16. Watanabe M, Sudo R (2019) Establishment of an in vitro vascular anastomosis model in a microfluidic device. J Biomech Sci Eng 14:18-00521

17. Yamamura N, Sudo R, Ikeda M, Tanishita K (2007) Effects of the mechanical properties of collagen gel on the in vitro formation of microvessel networks by endothelial cells. Tissue Eng 13:1443–1453

Chapter 7

Spatial Statistics-Based Image Analysis Methods for the Study of Vascular Morphogenesis

Diego Guidolin, Cinzia Tortorella, and Domenico Ribatti

Abstract

Several studies are available addressing the mechanisms of vascular morphogenesis in order to unravel how cooperative cell behavior can follow from the underlying, genetically regulated behavior of endothelial cells and from cell-to-cell and cell-to-extracellular matrix interactions. From the morphological standpoint several aspects of the process are of interest. They include the way the pattern of vessels fills the available tissue space and how the network grows during the angiogenic process, namely how a main trunk divides into smaller branches, and how branching occurs at different distances from the root point of a vascular tree. A third morphological aspect of interest concerns the spatial relationship between vessels and tissue cells able to secrete factors modulating endothelial cells self-organization, thus influencing vascular rearrangement.

In the present chapter image analysis methods allowing for a quantitative characterization of these morphological aspects will be detailed and discussed. They are almost based on concepts derived from the theoretical framework represented by spatial statistics.

Key words Area vasculosa, Endometrium, Angiogenesis, Morphometry, Sholl analysis, Spatial point patterns

1 Introduction

Blood vessel patterns are formed by the combined action of both deterministic and random processes [1, 2]. Vascular network formation proceeds along three main stages: (1) cell migration and early network formation; (2) network remodeling; and (3) differentiation in tubular structures and development of capillary networks, characterized by typical intercapillary distances ranging from 50 to 300 μm. During embryonic life, blood vessels first appear as the result of vasculogenesis, whereas remodeling of the primary vascular plexus occurs by angiogenesis [3, 4]. Vasculogenesis is the formation of capillaries from endothelial cells which differentiate from groups of mesodermal cells, leading to the establishment of a primitive vascular plexus [5]. The term angiogenesis, instead, is

Domenico Ribatti (ed.), *Vascular Morphogenesis: Methods and Protocols*, Methods in Molecular Biology, vol. 2206, https://doi.org/10.1007/978-1-0716-0916-3_7, © Springer Science+Business Media, LLC, part of Springer Nature 2021

applied to the formation of capillaries from preexisting vessels [6] and is based on endothelial sprouting or intussusceptive microvascular growth [7].

Several studies are available addressing the mechanisms of vascular morphogenesis in order to unravel how cooperative cell behavior can follow from the underlying, genetically regulated behavior of endothelial cells and from cell-to-cell and cell-to-extracellular matrix interactions (*see* [8–13] for examples). They suggested that nonlinear cell dynamics driven by angiogenic factors, cell adhesion and chemotaxis can provide models able to reproduce with good accuracy the formation of capillary networks, helping to understand which cell behaviors are essential to structure the vascular network.

From the morphological standpoint some aspects of the process deserve consideration. A first question concerns the way the pattern of vessels fills the available space. This feature, indeed, is of importance to get a better understanding of tissue perfusion and therefore of vascular network efficiency. A second question of interest is how the network grows during the angiogenic process. Two structural aspects can be considered in this regard, namely how a main trunk divides into smaller branches, and how the branching occurs at different distances from the root point of a vascular tree. A third morphological aspect of interest concerns the relationship between vessels and tissue cells secreting angiogenic cytokines and proteases, since their distribution can modulate endothelial cells self-organization and influence vascular rearrangement. This feature can be of particular relevance when pathological conditions are considered. In cancer, for instance, angiogenesis plays a significant role in tumor growth and metastasis [14]. Many studies, indeed, have demonstrated a correlation between increased microvascular density and metastatic disease, tumor size, grade, and poor prognosis [15]. The potential role of tissue resident cells (as, for instance, mast cells and macrophages) in tumor angiogenesis and progression [16] has been recently explored by some studies. They showed a spatial association between these cells, vessels, and parenchyma, allowing to speculate about their involvement in the development of the vascular support to tumor tissue [17, 18].

To devise morphometric methods aimed at providing a quantitative description of the abovementioned structural features of vascular morphogenesis, spatial statistics (*see* [19]) can represent a particularly important and useful theoretical framework.

The fundamental feature of spatial statistics is its concern with phenomena whose spatial location is either of intrinsic interest or contributes directly to a stochastic model for the phenomenon in question [20]. In other words, the objective of spatial statistics is not only to describe how things are distributed in space (e.g., random, clustered, uniform) but also to determine if spatial proximity is playing a causal role in the observed distribution [21]. The three main branches of spatial statistics concern spatially continuous processes, spatially discrete processes (lattices and areal unit data) and spatial point processes respectively [19].

A spatial point pattern is a set of locations, irregularly distributed within a designated region and presumed to have been generated by some form of stochastic mechanism. In most applications, the designated region is essentially planar (two-dimensional Euclidean space), but one-dimensional applications are also possible, and three-dimensional applications are becoming more common in conjunction with the development of more sophisticated three-dimensional scanning microscopes and imaging devices. This branch of spatial statistics is of particular interest in microanatomy [22] since cell distributions can be very well described as patterns of points representing the locations of the cells in a microscopic tissue section, which make possible the use of the tools provided by spatial statistics to analyze the features of their distribution also involving the concept of interactions between near-neighboring cells. The same strategies, however, could also represent a useful approach for the morphometric characterization of morphological features exhibited by vascular trees. Often, indeed, their topological structure and spatial distribution can be quite well characterized by exploiting the positions of their branching points in the host tissue [1, 23, 24].

Image analysis methods based on point pattern analysis concepts and aimed at quantitatively characterizing structural features of the vascular networks, how they fill the available tissue space and the level of spatial association between resident cells and vessels will be the focus of the present methodological discussion.

2   Materials

2.1   Software Tools

1. All the here described image analysis procedures will refer to and can be realized by the public domain image analysis package Fiji (https://imagej.net/Fiji/Downloads), a specific distribution of the NIH ImageJ software [25] which includes a set of particularly useful plugins. In principle, however, the procedures here outlined can be implemented in a variety of different platforms (*see* **Note 1**).

2. The analysis of some of the data obtained from the here described morphometric methods can be conveniently performed by using the public domain software R (https://www.r-project.org/), powered by the package SPATSTAT [26] implementing procedures for the analysis of spatial point patterns. Both can be downloaded from https://cran.r-project.org/mirrors.html (*see* **Note 2**).

3. General statistical and spreadsheet software applications are of help for conventional data processing tasks.

2.2 Analysis of Vessel Tree Growth and Distribution

1. Egg incubator.
2. Embryonated White Leghorn chick eggs.
3. Silicone egg holder.
4. Scalpel blade, needles, syringes, sharp-pointed dissection scissors.
5. Cellophane tape.
6. Digital photo camera. Images of the chick area vasculosa will be used to illustrate image analysis procedures aimed at characterizing the branching structure of the capillary tree and the way it fills the available space.

2.3 Analysis of Microvessel and Tissue Cell Codistribution

1. Paraffin-embedded tissue of normal and neoplastic endometrium.
2. Microscope slides, ethanol, poly-L-lysine, mounting medium.
3. Microtome.
4. Primary antibodies against tryptase (Novocastra, Leica Biosystems, Nussloch, Germany) and factor VIII (Dako, Agilent Technologies, Santa Clara, CA, USA).
5. Polymer/HRP secondary antibody using DAB as chromogen and polymer/alkaline phosphatase (AP) secondary antibody using Fast Red as chromogen (EnVision®, Dako, Agilent Technologies, Santa Clara, CA, USA).
6. Harris hematoxylin.
7. Light microscope. Images of the dual-labeled tissue sections, showing microvessels and mast cells as factor VIII- and tryptase-positive structures respectively, will be used to illustrate image analysis procedures aimed at characterizing the cell-to-vessel spatial relationship.

3   Methods

3.1 Imaging Chick Embryo Area Vasculosa

A detailed description of the techniques to manage embryonated eggs in order to observe the chick embryo area vasculosa can be found in a previous volume of this series (*see* [27]). Briefly:

1. Disinfect fertilized eggs with 70% ethanol and let them dry. Incubate under conditions of constant humidity at a temperature of 37 °C.
2. After 72 h saturate a stack of gauze with 70% ethanol and swab the egg. Place a piece of plastic tape on the part of the egg to be windowed.

3. By using the point of scissors or a scalpel blade, open a small squared window in the eggshell after removal with a syringe of 2–3 ml of albumen in order to detach the developing chorioallantoic membrane from the shell.

4. Close the opening with cellophane tape and continue incubation.

5. Acquire pictures of the area vitelline at the selected stages of development.

3.2 Imaging Mast Cells and Microvessels in Tissue Sections

The EnVision® DoubleStain System was used to perform sample preparation according to the associated protocol (*see* **Note 3**). Briefly:

1. Cover sections with 0.03% hydrogen peroxide containing sodium azide to block endogenous peroxidases. Incubate 5 min, then rinse gently with a buffer solution.

2. Apply enough anti-tryptase antibody diluted 1:200 and incubate overnight at 4 °C. Rinse slides and apply enough polymer/HRP secondary antibody. Incubate 30 min then rinse, apply the chromogen (DAB) containing solution for 5–15 min, then rinse slides again.

3. Incubate samples for 30 min with the second primary antibody against factor VIII diluted 1:250 and, after rinsing, apply the polymer/AP secondary antibody and the chromogen (Fast Red) containing solution as outlined before.

4. Optionally, counterstain with hematoxylin.

5. Coverslip slides and acquire pictures at the light microscope.

3.3 Morphometric Characterization of Vascular Branching

Previously used to quantify characteristics of dendritic processes branching off neuronal cell bodies, Sholl analysis [28, 29] examines the number of branches that intersect concentric circles of increasing radii around the central root point of the vascular tree (*see* Fig. 1). Thus, it provides a distribution of branching at regular distances around the root of the vascular tree. From this distribution at least two metrics can be derived:

- The *Sholl's regression coefficient* [30] can be estimated by the so-called semi-log method by calculating the function Υ $(r) = N/S$, where N is the number of branches crossing a circle of radius r and S is the area of that same circle. The following first order linear regression (linear fit) of the base 10 logarithm of this function is then performed:

$$\log_{10} \Upsilon(r) = -k \cdot r + m,$$

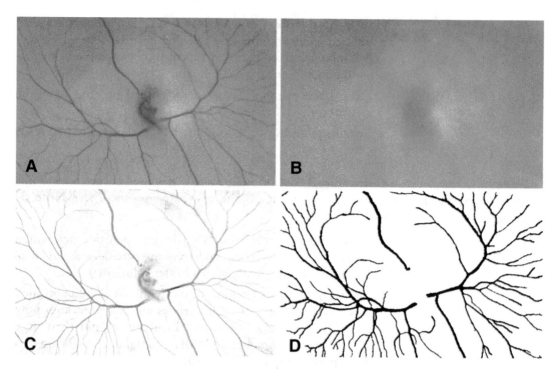

Fig. 1 Discrimination of the area vasculosa vascular tree by local thresholding. (**a**) Original image. (**b**) The background is estimated by applying to the original image a 30 × 30 pixels smoothing filter. (**c**) Image obtained following background subtraction: it virtually contains the vascular profiles only. (**d**) By applying to this image a conventional thresholding operation a binary image of the vascular tree can be obtained. If needed, procedures of interactive editing (by exploiting the set of filters and graphic tools available in Fiji) can be used to remove any artifact or image defect

where k is just the Sholl's regression coefficient. It is a measure of the change in density of branches as a function of distance from the root point and has been shown to have good discrimination value between various tree types.

- The *Branching index* (BI) [31] is defined as follows:

$$BI = \Sigma \, (\text{intersections circle}_n - \text{intersections circle}_{n-1})r_n.$$

If the number of intersections in the outer circle is equal or smaller than the number in the inner circle, the BI value is 0 indicating that the neurites have not ramified. Thus, the BI value is relative to the amount of branches that the vascular network produces. Furthermore, since the BI value also depends on the radius of the circles, it results higher when the vascular tree ramify far away from the root point than when it ramify in proximity of that central point (*see* **Note 4**).

The branching pattern is determined by the division of a main trunk into smaller branches. Bifurcation and trifurcation are the most commonly described modes, although other types have been observed [32]. The modality of vessel ramification, however, does not represent a feature that Sholl analysis can estimate and a specific procedure must be devised to get this information.

The whole image analysis approach is here outlined, using images of the chick embryo area vasculosa as an example [33] (*see* **Note 5**).

1. Start the Fiji software and load the image to analyze (*File → Open...*).

2. If needed, adjust contrast (*Image → Adjust → Brightness/ Contrast...*) and convert the image to 8-bit gray level (*Image → Type → 8-bit*).

3. An adaptive discrimination procedure [34] can then be applied to select vessel profiles virtually exclusive of all background. This method operates with a local threshold: the mean gray value of a neighboring region of suitable size is calculated for every pixel and this value plus an offset threshold constant defines the local threshold for each pixel. This operation can be realized (*see* **Note 6**) by first applying a background subtraction (*Process → Subtract background...*) and then a conventional thresholding operation (*Image → Adjust → Threshold...*). As a practical criterion, the size of the smoothing filter used to estimate the background should be in the order of the diameter of the vascular branches (in the example of Fig. 1 it was of 30×30 pixels). If needed, the resulting binary image can be refined by automatic and/or interactive editing procedures.

4. On the obtained binary image of the vascular tree Sholl analysis can be performed (Fig. 2) by executing a specific plugin (*see* [35]) available in Fiji (*Analyze → Sholl → Sholl analysis (from image)...*). A practical description of the plugin settings and usage can be found at https://imagej.net/Sholl_Analysis.

5. As shown in Fig. 2b, the main output of the plugin is a data table reporting the number of intersections at each sampled radial distance. From this basic measure the abovementioned metrics, namely *Sholl's regression coefficient* and *Branching index*, can be directly calculated by processing output data with general statistical and/or spreadsheet software applications.

6. Skeletonize the binary image of the vascular tree (*Process → Binary → Skeletonize*), then execute the plugin 'Analyze skeleton' [36] available in Fiji (*Analyze → Skeleton → Analyze skeleton*) with the 'display labeled skeletons' option active. A general description of the plugin can be found at https:// imagej.net/AnalyzeSkeleton. As shown in Fig.2c, d, the plugins provide as an output a new image of the skeleton where

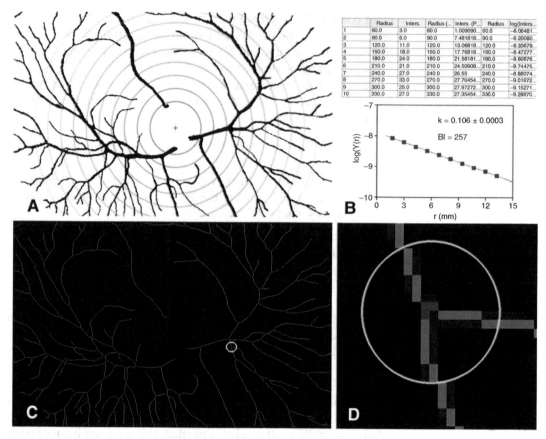

	Radius	Inters.	Radius (...	Inters. (P...	Radius	log(Inters...
1	60.0	3.0	60.0	1.009090...	60.0	−8.06481...
2	90.0	6.0	90.0	7.481818...	90.0	−8.20080...
3	120.0	11.0	120.0	13.06818...	120.0	−8.33679...
4	150.0	18.0	150.0	17.76818...	150.0	−8.47277...
5	180.0	24.0	180.0	21.58181...	180.0	−8.60876...
6	210.0	21.0	210.0	24.50909...	210.0	−8.74475...
7	240.0	27.0	240.0	26.55	240.0	−8.88074...
8	270.0	33.0	270.0	27.70454...	270.0	−9.01672...
9	300.0	25.0	300.0	27.97272...	300.0	−9.15271...
10	330.0	27.0	330.0	27.35454...	330.0	−9.28870...

$k = 0.106 \pm 0.0003$

$BI = 257$

Fig. 2 (**a**) Schematic description of the Sholl analysis. The vascular tree is sampled by a set of circles of increasing radius centered on the root point of the tree and the number of intersections of each circle with vascular branches is recorded. (**b**) The typical result table provided by the Fiji plugin performing this analysis is shown. From these data the two metrics described in Subheading 3.3, namely, the *Sholl's regression coefficient* and the *Branching index*, can be easily calculated. (**c**) Result of the skeletonization of the binary image of the vascular tree as provided by the "Analyze skeleton" plugin in Fiji: branches, branching points and terminal points are represented with different gray levels, allowing for a quite easy identification of branching points. (**d**) To assess the order of branching at each branching point a circular region of suitable size (yellow circle) is considered around each point and the number *n* of branches falling in it is estimated: the order of branching is *n − 1* (=2 in the illustrated example)

branches, branching points and end points are identified and tagged with different gray levels. From this image, therefore, two additional binary pictures can be immediately obtained containing only branching points and branches respectively. Global morphometric features such as the number of branching points and the number of branches can then be evaluated if needed.

7. The two abovementioned binary images can also represent the input of a procedure aimed at estimating the order of branching at each branching point. To evaluate this feature a circular region of suitable size (*see* Fig. 2d) centered on each branching point was explored to count the number *n* of branches falling in it. The number *n − 1* was then considered as the index of the branching mode occurring at the considered branching point [33]. A Fiji macro script to automate this process can be outlined as follows:

```
// General settings and definitions of the parameters

var t0="#Bin0"; //Identifiers of the binary images
var t1="#Bin1";


var maxdata=200; //Data buffer size
var idx=-1; //Buffer index


// Parameters:
var xp=newArray(maxdata); //x-coord of branching points
var yp=newArray(maxdata); //y-coord of branching points
var nb=newArray(maxdata); //N of branches per branching point
var ds=10; //Radius of the circular search area around each
branching
 point


 // After loading the two binary images get their identi-
fiers...

macro "Set image of branching points" {
  if (nImages==0)
  exit ("No image available...");
  t0=getTitle();
 }

 macro "Set image of branches" {
  if (nImages==0)
  exit ("No image available...");
  t1=getTitle();
 }

 // ...then perform measurement

 macro "Measure" {
  //First: from the image of branching points get their
 coordinates
  selectWindow(t0);
  run("Clear Results");
```

```
    run("Analyze Particles...", "size=0-Infinity circular-
ity=0.00-1.00 show=Nothing clear");
  BPoints=nResults;
  for(i=0; i&lt;BPoints; i++) {
  xp[i]=getResult('XM',i);
  yp[i]=getResult('YM',i);
  xp[i]=xp[i]-ds;
  yp[i]=yp[i]-ds;
  }
  //Second: in the image of branches analyze each branching
point location
  selectWindow(t1);
  sz=2*ds;
  for(i=0; i&lt;BPoints; i++) {
  makeOval(xp[i],yp[i],sz,sz); // draw a circular roi around
the branching point
    run("Analyze Particles...", "size=0-Infinity circular-
ity=0.00-1.00 show=Nothing clear");
  nb[i]=nResults; //number of branches falling in the roi
  nb[i]=nb[i]-1; //order of branching
  }
  run("Clear Results");
  for(i=0; i&lt;BPoints; i++) { //write the results table
  setResult("BrOrder", i, nb[i]);
  }
  updateResults(); //display results
  }
```

When loaded in Fiji (*Plugins → Macros → Install...*) the three functions defined by the macro script become available under *Plugins → Macros*.

3.4 Morphometric Characterization of the Vascular Tree Spatial Distribution

A structural feature of particular interest in order to get a better understanding of tissue perfusion, and therefore of vascular network efficiency, is the way the pattern of growing vessels fills the available space. To address this morphological feature spatial statistics methods using the distribution of the distances between branching points can be used for testing the randomness of the spatial pattern they generate and the deviations of the spatial distribution of points from randomness, such as local increase (clustering) or decrease (inhibition) of point density.

A spatial analysis method used to describe how point patterns occur over a given area of interest is the Ripley's K-function [37]:

$$K(t) = A \sum_i \sum_{j \neq i} w_{ij} \frac{I(d_{ij} < t)}{N^2}$$

where N is the number of points, A is the area of the study region, d_{ij} is the distance between the ith and jth points, and $I(x)$ is the indicator function with value 1 if x is true and 0 otherwise. The weight function w_{ij} provides edge correction: it has the value 1 if the circle centered on the point i and passing through the point j is completely inside the study area, while if part of the circumference falls outside the study area then w_{ij} is the proportion of the circumference of that circle falling in the study area. The K-function is generally calculated at multiple values of t in order to see how the point pattern distribution changes with scale. The observed K-function is usually compared to the one expected to find based on a completely spatially random point pattern of the same size: if the number of points found within a given distance of each individual point is greater than that for a random distribution, the experimental distribution is clustered, if the number is smaller, the distribution is dispersed. To estimate the K-function under conditions of spatial randomness, a popular strategy is represented by Monte Carlo methods (*see* [22]). This approach involves the simulation in the study area of a large number of random spatial point patterns of the same size as the observed point pattern. The mean K-function $(K_0(t))$ together with the upper and lower 95% envelopes $(K_{0L}(t), K_{0U}(t))$ of the K-functions for the simulated realizations of a completely random point process can then be calculated [38]. A convenient way to express the results is to apply the following transform [39] to all the obtained K-functions $(K(t), K_0(t), K_{0L}(t), K_{0U}(t))$:

$$H(t) = \sqrt{\frac{K(t)}{\pi}} - t$$

which transforms the theoretical random (Poisson) K-function to the straight line $H(t) = 0$. When the empirical H-function is plotted together with the lower and upper-envelope H-functions (*see* Fig. 3), t-intervals where the observed function is outside the range for a random point process become immediately evident. They correspond to point-to-point distances significantly different than expected by chance.

The main procedural steps to perform the just described analysis can be outlined as follows:

1. Obtain a binary image (Fig. 3a) of the branching points as described before (Subheading 3.3, **step 6**) and in Fiji select "center of mass" from *Analyze → Set measurements...*

2. Measure the coordinates of the branching points (*Analyze → Analyze particles...*) and save the results as a tab-delimited text file:

```
x y
1 373.864 1.409
```

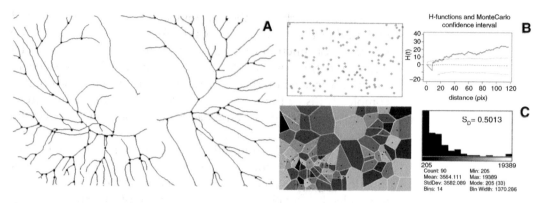

Fig. 3 (a) Skeleton of the vascular tree shown in Fig. 1a in which branching points are highlighted. The (*x,y*) coordinates of these points are the input data for procedures aimed at characterizing the way the pattern of growing vessels fills the available space. (**b**) The *H*-function (see text) of the pattern of branching points is shown together with the upper- and lower-95% envelopes for a random distribution. A set of random point patterns generated in the same area of interest (one representative example is shown) was used to estimate the envelopes (*see* subheading 3.4). In the illustrated example, the empirical *H*-function indicates that the vascular tree doesn't uniformly fill the tissue space but appears significantly aggregated since the *H*-function is outside the superior limit for a random point process for a large interval of point-to-point distances. (**c**) Voronoi diagram corresponding to the branching point distribution in A. and estimation of the parameter S_D (*see* subheading 3.4) from the mean area and standard deviation exhibited by the Voronoi's cells

```
2 503.827 9.118
3 328.821 17.286
4 313.232 36.915
5 362 57.500
6 ... ...
```

3. Open the software R and load the package SPATSTAT (*Packages → Install packages...*).

4. In order to automate the SPATSTAT-based point pattern analysis procedure, a R script file represents a useful tool. It can be outlined as follows:

```
require(spatstat, quietly=TRUE) #, save=FALSE) #declare that
SPATSTAT has to be used

pb <- list(x=c(0,639,639,0), y=c(0,0,479,479)) # define the
study area (a 640x480 image in
  this example)
p <- list(pb)
w <- owin(poly=p)

# Load the point pattern from file
#
coord <- read.table("path/filename.txt", header=TRUE)
```

```
pp <- ppp(x=coord$x, y=coord$y, window=w)
plot(pp)

# Calculate the empirical K-function
#
emp.K <- Kest(pp)
 range.r <- attr(emp.K, "alim")
 upper.r <- range.r[2]
 seq.r <- emp.K$r[emp.K$r<=upper.r]

# Monte Carlo simulation of the random point pattern
#
nsim <- 99 # No.of simulated patterns

av.K <- numeric(length(seq.r))
std.K <- numeric(length(seq.r))
U.K <- numeric(length(seq.r))
L.K <- numeric(length(seq.r))

for(i in 1:nsim) { # Perform the simulations
 print(i)
 pp0 <- rpoint(pp$n, 1, win=w)
 S.K <- Kest(pp0)
 km.K <- S.K$trans[S.K$r<=upper.r]
 av.K <- av.K + km.K
 std.K <- std.K+km.K*km.K
 }

# Calculate the mean K-function of the random process and 95%
envelopes
 #
 std.K <- sqrt((nsim*std.K-av.K^2)/(nsim*(nsim-1))) # Ko
standard deviation
 av.K <- av.K/nsim # mean Ko
 L.K <- av.K-1.96*std.K # 95% c.i.-lower
 U.K <- av.K+1.96*std.K # 95% c.i.-upper

# Calculate H-transforms
#
emp.H <- sqrt(emp.K$trans/pi)-emp.K$r
av.H <- sqrt(av.K/pi)-seq.r
L.H <- sqrt(L.K/pi)- seq.r
U.H <- sqrt(U.K/pi)-seq.r

# Output L-transforms
#
 plot(emp.K$r, emp.H, type="l", xlim=c(0,upper.r), ylim=c
(-20,40),xlab="distance",
```

```
    ylab="H(r)", main="H-functions and MonteCarlo confidence
interval")
lines(seq.r,av.H, lty=2)
lines(seq.r,U.H, lty=3)
lines(seq.r,L.H, lty=3)
```

5. The script can then be saved as a text file to be loaded in R (*File → Open...*) and runned (*Edit → Run...*). The result is a plot of the empirical *H*-function together with the confidence limits for a random point pattern of the same size as the observed one (Fig. 3b).

A more concise way to describe the heterogeneity in space occupancy of the set of branching points can be the estimation of some uniformity parameter, expressing the deviation from a random spatial distribution exhibited by the observed pattern of points. In this respect, Voronoi's analysis [40] was shown to be a useful tool [1]. Given a spatial point pattern the analysis starts from the calculation [41] of a diagram (Voronoi diagram), consisting in a tessellation of the plane with cells whose sides are segments bisecting the lines linking each point with its nearest neighbors. As shown in Fig. 3c, the structure of the diagram provides relevant information about the level of order/disorder exhibited by the spatial arrangement of the points. In particular, when heterogeneity increases, the following parameter (ranging between 0 and 1) also increases:

$$S_D = 1 - \left(1 + \frac{\sigma_A}{\bar{A}}\right)^{-1},$$

where $\bar{A}$ is the mean area of the Voronoi's cells and σ_A their standard deviation.

The main steps of this type of analysis can be summarized as follows:

1. Obtain a binary image of the branching points (Fig. 3a) as described before.

2. Execute the Fiji function *Process → Binary → Voronoi* to obtain the Voronoi diagram of the point pattern, then threshold the cells of the diagram (*Image → Adjust → Threshold...*).

3. From *Analyze → Set measurements...* select "*Area*", then *Analyze → Measure*. The list of the area values of each single Voronoi cell become visible in the "Results" window.

4. Executing *Results → Distribution...* the histogram of the area values will be estimated together with the mean area and the standard deviation. From these data S_D can be immediately calculated (*see* Fig. 3c).

To characterize the spatial relationship between cells and vessels, the same basic idea developed in the previous section can be used. In this case, however, input data will be the distances between pairs of events of different type (i.e., cells and vascular profiles), and a convenient metric is represented by the cumulative frequency distribution ($G(d)$) of all the observed cell-to-vessel distances [17]. Its expected value under complete spatial randomness ($G_0(d)$) can be estimated by averaging the cumulative frequency distributions of the distances obtained from a number of simulated random point patterns with the same characteristics as the observed one. To interpret the cell-to-vessel spatial relationship statistically, the 95% confidence envelope for $G_0(d)$ can also be calculated from the Monte Carlo simulations [42]. The null hypothesis is that there is no difference between the two functions, that is, $G(d) = G_0(d)$ for all d. Thus, if $G(d)$ is greater than the confidence envelope around $G_0(d)$, then the cells are clustered around the vessels, that is, they are closer to the vessels than expected by chance. If $G(d)$ is lower than the envelope around $G_0(d)$, then short cell-to-vessel distances are less frequent than expected by chance; that is, the placement of the cells close to the vessels was "inhibited." As discussed before (*see* Subheading 3.4), to make such a comparison easier the following transform can be applied to $G(d)$ and to the functions defining the envelope around $G_0(d)$:

$$L(d) = G(d) - G_0(d)$$

The analysis can be performed according to the following main steps referring to images of the dual-labeled tissue sections (*see* Subheading 2.2 and Fig. 4a) where microvessels and mast cells (MC) were imaged by immunohistochemistry [17]:

1. Start the Fiji software, load the image to analyze (*File → Open...*), and, if needed, adjust contrast (*Image → Adjust → Brightness/Contrast...*) and correct background unevenness (*Process → Subtract background...*).

2. Color deconvolution is then applied to allow the identification of vessels and MC profiles. This procedure implements stain separation according to the method by Ruifrok et al. [43] and can be performed by using a specific plugin implemented in Fiji (*Image → Color → Color deconvolution...*). Detailed information and a practical description of the plugin settings and usage can be found at https://blog.bham.ac.uk/intellimic/g-landini-software/colour-deconvolution/. As shown in Fig. 4b, this procedure allows the generation of two binary images containing Fast Red- and DAB-stained structures, respectively.

3. To estimate the distance of each MC profile from the vessels, the binary image of the vessel profiles was further processed to calculate its "distance transform" [34]. This transform can be

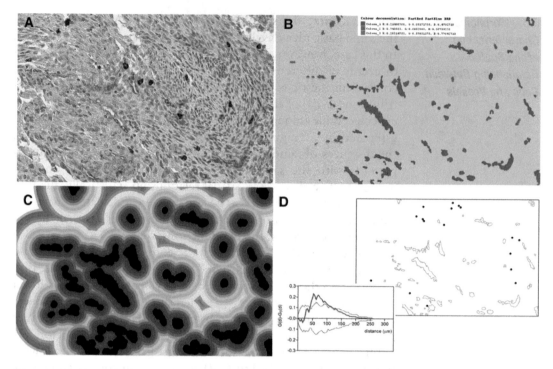

Fig. 4 Analysis of the cell-to-vessel spatial relationship. (**a**) Image of an immunohistochemical sample showing factor VIII-positive vessels (red) and tryptase-positive MC (brown). (**b**) By applying a color deconvolution procedure [43] the two stains can be efficiently separated. The procedure involves a transformation of the color space used to represent pixel colors from the conventional RGB space to a color space specific to the applied staining technique. The transforming factors for the histochemical technique here used are shown in the inset. (**c**) Distance transform of the image shown in B.: background pixels are labeled according to their distance from the nearest vessel profile boundary as indicated by the colored intervals. Thus, the distance of each cell profile from the vessels can be estimated by the value of the map at the point where the cell profile is located (i.e., at its x-, y-coordinates). (**d**) The *L*-function (see text) describing the cumulative distribution of cell-to-vessel distances is shown together with the upper- and lower-95% envelopes for a random distribution. A set of random point patterns generated in the same area of interest (one representative example is shown) was used to estimate the envelopes (see text). The empirical *L*-function shows that a range of distances occur with a frequency higher than expected by chance, indicating a significant spatial association between MC and vessels

obtained by executing a specific function available in Fiji (*Process → Binary → Distance Map*). The algorithm provides a map where each background pixel is labeled (*see* Fig. 4c) with a value equal to its distance from the nearest pixel belonging to a vessel profile.

4. The distance from vessels of each MC profile can then be estimated by the value the distance map exhibits at the location where the cell profile is located. To this end, from *Analyze → Set measurements. . .* select "*Mean Grey Value*" and "*Redirect to: <distance map name>*". Move to the binary image of cells and execute *Analyze → Analyze particles. . .* The result window will

display the list of distances of each cell from the nearest vascular profile:

```
dist
1 41.776
2 122.529
3 82.380
4 82.945
5 128.732
... ...
```

5. Generate a number n of random point patterns of the same size as the observed one around the vascular pattern under analysis (*see* Fig. 4d). To support the Monte Carlo simulation the development of a specific Fiji macro is of help. An example is provided by the following script:

```
var t0="#Img"; //Image identifiers
var t1="#csr";
var n=20;
var id0, id1;

// Set the size (number of "cells") of the point pattern

macro "Set size" {
 n=getNumber("Pattern size = ",n);
}

// Starting from the binary image of the vascular profiles
// a random point pattern is generated around them

macro "Get CSR" {
 t0=getTitle(); //Image of vascular profiles
 id0=getImageID();
 w=getWidth();
 h=getHeight();
 t1=t0+"-CSR";
 run("Duplicate...", "title=["+t1+"]"); //Image to host the
random pattern
 run("Select All");
 setBackgroundColor(0, 0, 0);
 run("Fill", "slice");
 id1=getImageID();
 i=0;
 while(i<n) { //Place n random points in the background
 r=random(); //surrounding vessels
 x=round(r*w);
 r=random();
 y=round(r*h);
```

```
selectImage(id0);
value=getPixel(x,y);
if (value==0) {
selectImage(id1);
setPixel(x,y,255);
i+=1;
}
}
}
```

When loaded in Fiji (*see* Subheading 3.3) the two defined functions will become available under *Plugins → Macros*, and each time they are executed a random point pattern is generated (*see* Fig. 4d) and the distances from vessels of the generated points can then be measured as described at **step 4** and stored for further analysis. Repeating the procedure *n* times will provide *n* estimates of a random point pattern.

6. Repeating **steps 1–5** for all the images corresponding to the microscope fields considered to sample the tissue will lead to the final set of (*n + 1*) lists corresponding to empirical and random cell-to-vessel distances.

7. By using a standard statistical software package, from each list of distances a cumulative distribution can be obtained, and by using a conventional spreadsheet software the mean cumulative distribution of the random distances $(G_0(d))$ together with their upper and lower 95% confidence envelopes can be calculated (*see* Subheading 3.4). Finally, the empirical cumulative distribution $(G(d))$ and the 95% envelopes around $G_0(d)$ can be transformed to the abovementioned *L*-function for a direct statistical comparison. A typical table of the final results, therefore, can appear as follows:

| d (microns) | Empirical cumulative distribution $G(d)$ (rel. freq.) | Cumulative random distributions | | | Mean cumulative random distribution | | Empirical L transform | Upper 95% envelope | Lower 95% envelope |
		Random 1 (Rel. freq.)	Random 2 (rel. freq.)	Random n (rel. freq.)	$G_0(d)$ (rel. freq)	std	$L(d) = G$ $(d) - G_0(d)$	$1.96 \times$ std	$-1.96 \times$ std
0.00	0.02	0.02	0.02	... 0.04	0.02	0.02	0.00	0.04	−0.04
4.16	0.04	0.08	0.06	... 0.10	0.05	0.03	−0.01	0.06	−0.06
8.32	0.08	0.13	0.10	... 0.16	0.11	0.05	−0.03	0.10	−0.10
12.48	0.18	0.13	0.23	... 0.18	0.15	0.06	0.03	0.11	−0.11
...	...	...	...	... ...	...	...	...	...	...

When the L-function is plotted together with the upper- and lower-95% envelopes, the distance intervals where the *L*-function is outside the range for a random point process will become immediately evident (*see* Fig. 4d).

---

4   Notes

1. The described image-analysis procedures can also be performed by using the standard release of ImageJ (https://imagej.nih.gov/ij/download.html). In this case, the mentioned plugins have to be downloaded from the ImageJ plugins repository or from other sources and manually installed. The basic input data to most of the procedures here proposed to quantitatively describe a vascular network structure are the coordinates of the branching points of the vascular tree. In this respect, they are also a main output provided by presently available integrated software packages for the image analysis of vascular networks, such as "AutoTube" [44] and "AngioQuant" [45].

2. A public domain software useful to manage R scripts and procedures is "Rstudio" (https://rstudio.com/products/rstudio/download/). It provides a set of integrated tools (such as a console, a syntax-highlighting text editor and tools for plotting) to assist in the production of R-based applications.

3. The image analysis procedures aimed at characterizing cell distributions around vessels (Subheading 3.5) are not restricted to the immunohistochemical protocol applied as an example in Subheading 3.2. Thus, samples can be prepared according to different immunohistochemical protocols, provided the result is a double staining differentiating cells of interest and vascular profiles. Of course, some of the parameters involved in the analysis (in particular those discussed at **step 2** of Subheading 3.5) would need some adjustment depending on the histochemical approach used.

4. It is important to stress that since the BI is directly proportional to the radius, in order to calculate the BI using the Sholl analysis the distance between circles must be consistently maintained in the experiment when comparing different vascular networks (*see* [33]). As a rule, the analysis with little separation between circles produces a more refined value of the BI [31].

5. The described analysis can also be applied (with minor changes) to other 2-dimensional vascular networks, as, for instance, the chorioallantoic membrane [1, 46] and the vascular network of the retina [47].

6. The proposed procedure to realize a local thresholding is the most common one and allows user interaction. More automatic procedures to perform local thresholding can be found in Fiji under *Image → Adjust → Auto Local Threshold*. Depending on the images under examination, they could represent alternative efficient methods.

Acknowledgments

This work has been supported by University of Padova, grant DOR2018.

References

1. Guidolin D, Nico B, Mazzocchi G, Vacca A, Nussdorfer GG, Ribatti D (2004) Order and disorder in the vascular network. Leukemia 18:1745–1750

2. Guidolin D, Crivellato E, Ribatti D (2011) The "self-similarity logic" applied to the development of the vascular system. Dev Biol 351:156–162

3. Patan S (2000) Vasculogenesis and Angiogenesis as mechanisms of vascular network formation, growth and remodeling. J Neurooncol 50:1–15

4. Ribatti D, Nico B, Crivellato E (2015) The development of the vascular system: a historical overview. In: Ribatti D (ed) Vascular morphogenesis: methods and protocols, Methods in molecular biology, vol 1214, pp 1–14

5. Risau W, Flamme I (1995) Vasculogenesis. Annu Rev Cell Dev Biol 11:73–91

6. Risau W (1997) Mechanisms of angiogenesis. Nature 386:671–674

7. Djonov V, Schmid M, Tschanz SA, Burri PH (2000) Intussusceptive angiogenesis: its role in embryonic vascular network formation. Circ Res 86:286–292

8. Vernon RB, Angello JC, Iruela-Arispe ML, Lane TF, Sage EH (1992) Reorganization of basement membrane matrices by cellular traction promotes the formation of cellular networks in vitro. Lab Invest 66:536–547

9. Manoussaki D, Lubkin SR, Vernon RB, Murray JD (1996) A mechanical model for the formation of vascular networks in vitro. Acta Biotheorel 44:271–282

10. Serini G, Ambrosi D, Giraudo E, Gamba A, Preziosi L, Bussolino F (2003) Modeling the early stages of vascular network assembly. EMBO J 22:1171–1179

11. Guidolin D, Rebuffat P, Albertin G (2011) Cell-oriented modeling of angiogenesis. ScientificWorldJournal 11:1735–1748

12. Daub JT, Merks RMH (2015) Cell-based computational modeling of vascular morphogenesis using *Tissue Simulation Toolkit*. In: Ribatti D (ed) Vascular morphogenesis: methods and protocols, Methods in molecular biology, vol 1214, pp 67–127

13. Guidolin D, Fede C, Albertin G, De Caro R (2015) Investigating in vitro angiogenesis by computer-assisted image analysis and computational simulation. In: Ribatti D (ed) Vascular morphogenesis: methods and protocols, Methods in molecular biology, vol 1214, pp 197–214

14. Hanahan D, Folkman J (1996) Patterns and emerging mechanisms of the angiogenic switch during tumorigenesis. Cell 86:353–364

15. Ribatti D, Nico B, Ruggieri S, Tamma R, Simone G, Mangia A (2016) Angiogenesis and antiangiogenesis in triple-negative breast cancer. Transl Oncol 9:453–457

16. Ribatti D (2013) Mast cells and macrophages exert beneficial and detrimental effects on tumor progression and angiogenesis. Immunol Lett 152:83–88

17. Guidolin D, Marinaccio C, Tortorella C, Annese T, Ruggieri S, Finato N, Crivellato E, Ribatti D (2017) Non-random spatial relationships between mast cells and microvessels in human endometrial carcinoma. Clin Exp Med 17:71–77

18. Guidolin D, Ruggieri S, Annese T, Tortorella C, Marzullo A, Ribatti D (2017) Spatial distribution of mast cells around vessels and glands in human gastric carcinoma. Clin Exp Med 17:531–539

19. Gelfand AE, Diggle PJ, Fuentes M, Guttorp P (eds) (2010) Handbook of spatial statistics. CRC Press, Boca Raton, FL

20. Diggle PJ (2010) Historical introduction. In: Gelfand AE, Diggle PJ, Fuentes M, Guttorp P (eds) Handbook of spatial statistics. CRC Press, Boca Raton, FL, p 3

21. Davis FW (1993) Introduction to spatial statistics. In: Levin SA, Powell TM, Steele JH (eds) Patch dynamics, Lecture notes in biomathematics, vol 96, pp 16–26

22. Diggle PJ (ed) (2014) Statistical analysis of spatial and spatio-temporal point patterns, 3rd edn. CRC Press, Boca Raton, FL

23. Zudaire E, Gambardella L, Kurcz C, Vermeren S (2011) A computational tool for quantitative analysis of vascular networks. PLoS One 6: e27385

24. Belle J, Ysasi A, Bennett RD, Filipovic N, Nejad MI, Trumper DL, Ackermann R, Wagner W, Tsuda A, Konerding MA, Mentzer SJ (2014) Stretch-induced intussuceptive and sprouting angiogenesis in the chick chorioallantoic membrane. Microvasc Res 95:60–67

25. Schindelin J, Rueden CT, Hiner MC, Eliceiri KW (2015) The ImageJ ecosystem: an open platform for biomedical image analysis. Mol Reprod Dev 82:518–529

26. Baddeley A, Turner R (2005) Spatstat: an R package for analyzing spatial point patterns. J Stat Softw 12(6):1–42

27. Makanya AN, Styp-Rekowska B, Dimova I, Djonov V (2015) Avian area vasculosa and CAM as rapid in vivo pro-angiogenic and anti-angiogenic models. In: Ribatti D (ed) Vascular morphogenesis: methods and protocols, Methods in molecular biology, vol 1214, pp 185–196

28. Sholl DA (1953) Dendritic organization in the neurons of the visual and motor cortices of the cat. J Anat 87:387–406

29. Sholl DA, Uttley AM (1953) Pattern discrimination and the visual cortex. Nature 171:387–388

30. Milošević NT, Ristanović D (2007) The Sholl analysis of neuronal cell images: semi-log or log-log method? J Theor Biol 245:130–140

31. Garcia-Segura LM, Perez-Marquez J (2014) A new mathematical function to evaluate neuronal morphology using the Sholl analysis. J Neurosci Methods 226:103–109

32. Kopylova VS et al (2017) Fundamental principles of vascular network topology. Biochem Soc Trans 45:839–844

33. Guidolin D, Tamma R, Tortorella C, Annese T, Ruggieri S, Marzullo A, Ribatti D (2020) Morphometric analysis of the branching of the vascular tree in the chick embryo area vasculosa. Microvasc Res 128:103935

34. Russ JC (1995) The image processing handbook. CRC Press, Boca Raton, FL, p 272

35. Ferreira T, Blackman A, Oyrer J, Jayabal A, Chung A, Watt A, Sjöström J, van Meyel D (2014) Neuronal morphometry directly from bitmap images. Nat Methods 11:982–984

36. Arganda-Carreras I, Fernandez-Gonzales R, Munoz-Barrutia A, Ortiz-De Solorzano C (2010) 3D reconstruction of histological sections: application to mammary gland tissue. Microsc Res Tech 73:1019–1029

37. Ripley BD (1981) Spatial statistics. Wiley, New York, NY

38. Smith TE (2016) Notebook on spatial data analysis. http://www.seas.upenn.edu/~ese502/#notebook

39. Kiskowski MA, Hancock JF, Kenworthy AK (2009) On the use of Ripley's K-function and its derivatives to analyze domain size. Biophys J 97:1095–1103

40. Seul M, O'Gorman L, Sammon MJ (2000) Practical algorithms for image analysis. Cambridge University Press, Cambridge

41. Fortune SJ (1987) A sweepline algorithm for Voronoi diagrams. Algorithmica 2:153–174

42. Gatrell AC, Bailey TC, Diggle PJ, Rowlingson BS (1996) Spatial point pattern analysis and its application in geographical epidemiology. Trans Inst Br Geogr 21:256

43. Ruifrok AC, Katz RL, Johnston DA (2003) Comparison of quantification of histochemical staining by hue-saturation-intensity (HSI) transformation and color-deconvolution. Appl Immunohistochem Mol Morphol 11:85–91

44. Montoya-Zegarra JA, Russo E, Runge P, Jadhav M, Willrodt A-H, Stoma S, Norrelykke SF, Detmar M, Halin C (2018) AutoTube: a novel software for the automated morphometric analysis of vascular networks in tissues. Angiogenesis 22:223–236

45. Niemistö A, Dunmire V, Yli-Harja O, Zhang W, Shmulevich I (2005) Robust quantification of in vitro angiogenesis through image analysis. IEEE Trans Med Imaging 24:549–553

46. Ribatti D, Guidolin D, Conconi MT, Nico B, Baiguera S, Parnigotto PP, Vacca A, Nussdorfer GG (2003) Vinblastine inhibits the angiogenic response induced by adrenomedullin in vivo and in vitro. Oncogene 22:6458–6461

47. Shen W-Y, Lai CM, Graham CE, Binz N, Lai YKY, Eade J, Guidolin D, Ribatti D, Dunlop SA, Racoczy PE (2006) Long-term global retinal microvascular changes in a transgenic vascular endothelial growth factor mouse model. Diabetologia 49:1690–1701

Chapter 8

Studying Angiogenesis in the Rabbit Corneal Pocket Assay

Lucia Morbidelli, Valerio Ciccone, and Marina Ziche

Abstract

The rabbit corneal micropocket assay uses the avascular cornea as a substrate to study angiogenesis in vivo. The continuous monitoring of neovascular growth in the same animal allows for the evaluation of drugs acting as suppressors or stimulators of angiogenesis. Through the use of standardized slow-release pellets, a predictable angiogenic response can be quantified over the course of 1–2 weeks. Uniform slow-release pellets are prepared by mixing purified angiogenic growth factors such as basic fibroblast growth factor (FGF) or vascular endothelial growth factor (VEGF) and a synthetic polymer to allow for their slow release. A micropocket is surgically created in the cornea thickness under anesthesia and in sterile conditions. The angiogenesis stimulus (growth factor but also tissue fragment or cell suspension) is placed into the pocket in order to induce vascular outgrowth from the limbal capillaries where vessels are preexisting. On the following days, the neovascular development and progression are measured and qualified using a slit lamp, as well as the concomitant vascular phenotype or inflammatory features. The results of the assay allow to assess the ability of potential therapeutic molecules to modulate angiogenesis in vivo, both when released locally or given by ocular formulations or through systemic treatment. In this chapter the experimental details of the avascular rabbit cornea assay, the technical challenges, advantages, and limitations are discussed.

Key words Angiogenesis, Capillary, Endothelial cell, Angiogenic factors, Angiogenesis inhibitors, Drug treatment, Slow release

1 Introduction

In order to develop and evaluate drugs acting as suppressors or stimulators of angiogenesis, the continuous in vivo monitoring of angiogenesis is required. In this respect, there is the continuous work to provide preclinical models for quantitative analysis of in vivo angiogenesis [1, 2]. The cornea assay consists in the placement of an angiogenesis inducer (tumor tissue, cell suspension, growth factor) into a micropocket made in the cornea thickness and the evaluation of vascular outgrowth from the peripherally located limbal vessels toward the stimulus. Since the rabbit cornea is initially avascular, this assay has the advantage of measuring only new blood vessels. The micropocket assay is, indeed, a suitable

Domenico Ribatti (ed.), *Vascular Morphogenesis: Methods and Protocols*, Methods in Molecular Biology, vol. 2206,
https://doi.org/10.1007/978-1-0716-0916-3_8, © Springer Science+Business Media, LLC, part of Springer Nature 2021

model to characterize the pro- and antiangiogenic effect of modulators and drugs with potential application in other fields as wound healing or cancer or ocular disease.

Different antiangiogenic molecules have been found in the cornea as angiostatin, endostatin, interleukin-1 receptor antagonist, pigment epithelium-derived factor, and thrombospondin [3–5]. Recently, the preservation of the avascular phenotype of the cornea has been associated to high levels of soluble vascular endothelial growth factor receptor (sVEGR1), able to neutralize the VEGF-A present in the cornea [6]. Thus, vascularization occurring during different pathophysiological conditions is the result of the perturbed balance among redundant inhibitory mechanisms. Specifically, a series of diseases are associated with corneal vascularization as traumatic injuries, corneal graft rejection, infections, and chronic inflammation.

Folkman's group firstly described the corneal assay in New Zealand white rabbits in the early 1970s [7]. The assay was chosen for the absence of a preexisting vascular pattern in New Zealand white rabbits and for the easy manipulation of the cornea and continuous monitoring of the neovascular growth. Our group settled a series of modifications of the original method [7], having set up protocols for the implant of multiple samples, including cell suspensions and tissue fragments. This technique, extensively used during the years, has been substantially modified to characterize angiogenesis inducers, to validate angiogenesis inhibitors, to study the interaction between different factors and the cellular, biochemical, and molecular mechanism of angiogenesis.

Refinement of drug formulation for local delivery in the eye and pharmacokinetic profile in ocular tissue components have been established.

In the following sections the experimental details and protocols of the avascular cornea assay are presented.

1.1 Rabbit Cornea Pocket Assay

First of all, the protocols and treatments must be approved by the local laboratory animal ethical board and national agencies, according to current laws and guidelines (European Directive 2010/63/EU or ARVO Statement for the Use of Animals in Ophthalmic and Vision Research, and 3R guidelines as in http://www.nc3rs.org.uk/), since the surgical procedure requires general anesthesia and a moderate degree of discomfort.

The micropocket assay is performed in albino rabbits (*see* **Note 1**) and requires the simultaneous presence of two qualified operators during all the steps (*see* **Note 2**).

2 Materials

2.1 Animals

Specific-pathogen-free (SPF) New Zealand albino rabbits (Charles River, www.criver.com) of 1.5–2.5 kg (*see* **Note 3**).

2.2 Reagents and Drugs

1. Recombinant growth factors or drugs to be studied as slow-release preparation must be dissolved in water or phosphate buffered saline (PBS) or ethanol or methanol in highly concentrated solutions (0.1–1 mg/ml) (*see* **Note 4**).

2. Slow-release polymer. Different polymers can be used (*see* **Note 5** for a comparison with other polymers). Ethylene-vinyl acetate copolymer (Elvax 40) (DuPont de Nemours, Wilmington, DE, www.dupont.com) should be previously prepared and tested for biocompatibility [8] (**Note 6**).

3. 20 mg/ml xylazine solution (Xilor®, BIO 98 Srl, Milan, Italy).

4. Zoletil-20, a combination of a dissociative anesthetic agent, tiletamine hydrochloride, and a tranquilizer, zolazepam hydrochloride (each at 10 mg/ml) (Zoletil® Virbac Srl, Milan, Italy).

5. The local anesthetic 0.4% solution benoxinate or oxybuprocaine chlorohydrate.

6. Tanax (T-61), a euthanasic mixture containing 200 mg/ml embutramide, 50 mg/ml mebenzonium iodide, and 5 mg/ml tetracaine hydrochloride.

7. Fixative: 4% paraformaldehyde in PBS, pH 7.4.

8. Liquid nitrogen, isopentane, and O.C.T. Tissue-Tek medium or similar.

2.3 Facilities, Equipment, and Materials

1. Cell culture facility equipped with vertical laminal flow hood, stereomicroscope and autoclave.

2. Animal facility equipped with a sterile surgical room.

3. Disposable scalpels for ocular microsurgery (no. 10–11, Aesculap).

4. Sterile forceps, silver spatula, microsurgery scissors, micro spatula.

5. Teflon plates (10 × 10 cm) and 6 cm glass petri dishes.

6. Vacuum.

7. Latex dental dam for endodontic procedures (DentalTrey, www.dentaltrey.com).

8. Insulin syringes.

9. Slit lamp stereomicroscope equipped with a digital camera.

10. Histology equipment and materials.

3   Methods

3.1   Sample Preparation

The material under test can be in the form of slow-release pellets incorporating recombinant growth factors, cell suspensions, or tissue samples.

– Preparation of slow-release pellets: In order to be implanted in the cornea, angiogenic factors (i.e., VEGF, FGF-2, cytokines, or other molecules) have to be prepared in a semisolid state, enabling surgical implantation and gradual release of the factor from the polymer. Pellets (implants) bearing molecules to be tested are prepared under a laminar flow hood according to the following steps. A given amount of the compound to be tested is previously dried on a flat Teflon surface. Then, a predetermined volume of polymer casting solution (10 μl/pellet) is mixed with the dried compound on the Teflon plate by the use of a stainless steel spatula. After drying, the film sequestering the compound is cut into $1 \times 1 \times 0.5$ mm homogeneous pieces under a stereomicroscope by the use of Vannas scissors and Dumont n. 5 tweezers. The pellets (in open glass Petri dishes) are left under vacuum at 4 °C overnight to remove residual solvent. Empty pellets of polymer are used as negative controls, while, depending on the experimental design, VEGF or FGF-2-containing pellets are used as positive controls (*see* **Note 7**).

– When testing the corelease of different molecules from the same pellet, the two substances are let to dry closely in the Teflon plate and then incorporated in the same polymer preparation (*see* **Note 8**).

– Preparation of cell suspension: The intrinsic angiogenic potential due to different stages of tumor progression or to the expression of genes or gene products have been documented by our group as well as by others [9–12]. Prepare a cell suspension by trypsinization of confluent cell monolayers to a final dilution of $2–5 \times 10^5$ cells in 5 μl. When using cells, angiogenic response can be graded based also on the number of cells implanted into the corneal stroma.

– Preparation of tissue samples: Tissue samples of animal and human origin have been successfully implanted into the rabbit cornea to produce angiogenesis [13–16]. When tissues are tested, fragments are removed within 2 h from patients or animals and kept at 4 °C in complete medium. Fragments of 2–3 mg are obtained by cutting the fresh tissue samples under sterile conditions by the use of microdissection instruments under a stereomicroscope.

3.2   Surgery

1. Anesthetize animals with Xilor (0.5 ml, i.m.) followed by Zole-til (5 mg/kg i.m.) or alternatively sodium pentothal (10 mg/kg, i.v.) The deepness of anesthesia is checked as reflex to pressure (*see* **Note 9**).

2. Each eye is protruded by the use of a dental dam and a local anesthetic (0.4% benoxinate) is instilled just before surgery.

3. The pellet implantation procedure starts with a linear intras-tromal incision, parallel to the corneoscleral limbus (linear keratotomy), using a surgical blade (disposable scalpel n. 10). The corneal micropocket for the pellet implant is produced with a 1.5 mm pliable silver spatula with smooth edge blade in the lower half of the cornea (*see* **Note 10**).

4. Pellet implant: The implant is introduced through the keratot-omy line, parallel to the corneal epithelium and under it, in the external third of the stroma, up to 2 mm from the limbus. One single pellet is selected from the Petri dish using Dumont n. 5 tweezers and then introduced in the corneal pocket. Microfor-ceps are used to keep open the edge of the cut. Locate the implant at 2 mm from the limbus to avoid false positives due to mechanical stress and to favor the gradient diffusion of test substances in the tissue, toward the endothelial cells of the limbal plexus.

5. When two factors are tested simultaneously, make two inde-pendent and parallel micropockets (*see* **Note 11** for different protocols).

6. Cell or tissue implant: The pocket is produced with an enlarged base (4 mm) to allocate the sample. To reduce corneal tension before cell or tissue implant, a small amount (20–50 µl) of the aqueous humor can be drained from the anterior chamber with an insulin syringe.

7. By using a micropipette introduce 5 µl containing $2–5 \times 10^5$ cells in medium supplemented with 10% serum in the corneal micropocket. When the overexpression of growth factors/inhi-bitors by stable transfection of specific cDNA is studied, one eye is implanted with transfected cells and the other with the wild type or vector transduced cell line. Suitable cell lines for these experiments are mammary carcinoma cells (MCF-7) [17], lymphoma Burkitt's cells (DG75) [10], Chinese hamster ovary cells (CHO) [12]. It might be necessary to evaluate the angiogenic potential of drug-treated cells. In these experiments cell monolayers are pharmacologically treated before the implant (18–24 h). One eye is implanted with treated cells and the contralateral with control cells [17, 18].

8. Tissue fragments are inserted in the corneal pocket with the aid of Dumont n. 5 tweezers. The angiogenic activity of tumor samples is compared with healthy tissue [15].

3.3 Quantification of Neovascular Growth

1. Subsequent daily observation of the implants is made with a slit lamp stereomicroscope without anesthesia. The clinical evolution of the implants and of the ocular lesions are recorded, and the presence of corneal reactions, such as redness, corneal edema and the intensity of the corneal cellular infiltrate and the total area of neovascularization are scored. The use of slit lamp stereomicroscope on awake animals allows for the observation of newly formed vessels during time with prolonged monitoring, up to 1 month. When studying tumor induced angiogenesis, neovascularization can accompany tumor cell growth, as in [15].

2. An angiogenic response is scored positive when budding of vessels from the limbal plexus occurs after 3–4 days and capillaries progress to reach the implanted pellet in 7–10 days. Implants that fail to produce a neovascular growth within 10 days are considered negative, while implants showing an inflammatory reaction are discarded and animals treated to reduce pain.

3. During each observation the number of positive implants over the total implants performed is scored.

4. The potency of angiogenic activity is evaluated on the basis of the number and growth rate of newly formed capillaries, and an angiogenic score is calculated by the formula [vessel density × distance from limbus] [17, 19]. A density value of 1 corresponds to 0–25 vessels per cornea, 2 25–50, 3 50–75, 4 75–100, and 5 more than 100 vessels. The distance from the limbus is graded (in mm) with the aid of an ocular grid.

5. To understand the mechanism of progression and/or regression by drug treatment, the two parameters (density and length) are considered separately, thus documenting the activity of treatment on endothelial cell proliferation (density) respect to elongation and organization (length).

6. The anterior ocular pole images are computer analyzed at fixed times on animals under anesthesia. An advanced video camera connected to a color video monitor and a computer with video-bluster and special capture software are used to record corneal responses. In order to extract the vascular tree from every image, the following graphic steps are required:

   - Adjustment of contrast and brightness, in order to highlight the vascular tree (image conversion in a gray scale format can be helpful in this stage).

   - Image extraction of the vascular tree from the background.

   Commercially available software (i.e., Corel Photo Paint and Corel Draw; Adobe Photoshop and National Institute of Health Image J1.38×) can be used for these purposes [20].

3.4 Histological Examination and Immunohistochemical Analysis

Depending on the experimental design, histological or immuno-histochemical analysis of corneal sections can be performed at fixed times during angiogenesis progression or at the end of the observations [17].

1. Animals are sacrificed with intravenous injection of 0.5 ml of Tanax or sodium pentothal (30 mg/kg as bolus).

2. The corneas are removed, oriented, and marked (*see* **Note 12**), immediately frozen in isopentane cooled in liquid nitrogen for 10 s, and stored at −80 °C in O.C.T. Tissue-Tek medium. If required, the cornea can be fixed it in paraformaldehyde.

3. Seven-micrometer-thick cryostat sections are stained with hematoxylin and eosin and adjacent sections can be used for immunohistochemical staining (*see* **Note 13**).

3.5 Gene and Protein Expression

At fixed times or at the end of the experiment, corneas can be removed and snap-frozen in liquid nitrogen. By using standard extraction reagents and buffers, mRNAs and proteins can be isolated to assess gene transcription and protein expression during vascular and tissue responses or following drug treatment.

3.6 Drug Treatments and Pharmacokinetic Studies

When performing drug treatments for ocular pathologies and to validate stimuli or signaling pathway, different approaches can be followed.

1. Eye drops: isotonic buffers (i.e., PBS without calcium and magnesium) at physiological pH can be used to dissolve drugs to be studied for their ability to modulate corneal angiogenesis. Depending on drug nature and half-life, eye drop treatments can be performed twice-fivefold a day, soon after an angiogenic stimulus has been implanted in the cornea stroma. Awake animals are immobilized in appropriate contention boxes. By the use of a sterile pipette, 50–100 μl of the drug solution is put in the subconjunctival space by pulling the lower lid. The eye is then kept close for at least 30 s to avoid liquid dispersion and dropout. The results about UPARANT eye drop treatment on VEGF-induced angiogenesis is reported in Fig. 1.

2. Ointment and gels: Simple eye ointment contains liquid paraffin (mineral oil) and wool fat (lanolin) in a yellow soft paraffin base (*see* **Note 14**). These ingredients produce a transparent, lubricating, and moistening film on the surface of the eyeball. Drug mixing is performed under hood and insulin syringes are prepared. 100 μl of ointment are poured in the subconjunctival space once or twice a day. Eye lids are closed and gently kneaded to form a film of the ointment or gel on the eye surface [22].

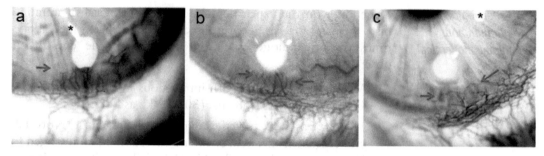

Fig. 1 Modulation of VEGF induced angiogenic response in rabbit eyes treated with eye drops containing UPARANT [21]. Angiogenesis by VEGF implants in eyes treated with vehicle, 100 and 500 µg PBS-UPARANT given five times/day every day from day 0 (4 h after implantation) until day 8. All eyes received corneal implants of polymer Elvax 40 pellet containing 200 ng of VEGF. Representative pictures of VEGF induced vascularization following treatment with vehicle (**a**), UPARANT 100 µg/dose (**b**) and 500 µg/dose (**c**), taken at day 10. Arrows indicate newly formed vessels. Asterisks represent flash artifact

3. Intravitreal injections (30–50 µl/eye) can be also performed under general anesthesia to study drug stability in the vitreous and diffusion to retina or to the anterior chamber, to obtain data closely related to human ocular pharmacology.

4. When drugs or genes transduced by viral vectors have to be locally tested, microinjection of concentrated solutions is performed by the use of insulin syringes equipped with 30 G needles. After the removal of aqueous humor, a volume of 10 µl is injected within the corneal stroma in the space between the limbus and the pellet implant [23].

5. At fixed times after treatment started, after animal sacrifice, all the eye tissues (cornea, aqueous humor, lens, vitreous humor, retina) can be isolated and frozen in liquid nitrogen, and tissue homogenates assessed for drug distribution and metabolism.

3.7 Advantages and Limitations of Rabbit Cornea Assay

1. Respect to the use of smaller animals (mouse, rat), the rabbit has a series of advantages. Due to the easy manipulation of rabbit cornea, not only purified growth factors/drugs are studied but also tumor tissue samples and cell suspensions. Since the rabbits are amenable and do not require anesthesia for daily monitoring, the continuous observation of neovascular growth in the same animal allows for the evaluation of drugs acting as suppressors or stimulators of angiogenesis, with a significant reduction in the total number of animals needed for statistical assessment.

2. The rabbit size (1.8–2.5 kg) lets an easy manipulation of the animal; the eye may be easily extruded from its location for surgery manipulation and daily observation.

3. Rabbit cornea has been found avascular in all strains examined so far. Rabbits are docile and amenable to handling and experimentation than mice and rats. In case of inflammatory reactions, these are easily detectable in rabbits by stereomicroscopic examination as corneal opacity, edema, and swelling.

4. Multiple observations are easily performed in rabbits, thus reducing the number of animals required for statistical evaluation. The use of slit lamp stereomicroscope and of awake animals allows for the observation of newly formed vessels during time with prolonged monitoring, up to 1 month. Histological analysis of the tissue allows for the characterization of the molecular and cellular events associated with the measured response.

5. Interestingly, corneal assay can also be used to study other biological processes. For example, studies of lymphangiogenesis were made possible through the implantation of low-dose FGF-2 pellets, which allowed for the visualization of lymphatic vessels through specific molecular markers [24].

6. In the rabbit eye, due to its wide area, stimuli in different forms can be introduced. In particular, the activity of specific growth factors can be studied in the form of slow-release pellets [10, 25–27] and of tumor or non-tumor cell lines stably transfected for the overexpression of angiogenic factors [10, 17, 18]. Cells with double transfection can also be studied [10, 11]. The implant of tumor samples from different locations can be performed both in corneal micropockets and in the anterior chamber of the eye to monitor angiogenesis produced by hormone-dependent tissues or tumors (i.e., human breast or ovary carcinoma in female rabbits), allowing for the detection of both the iris and the corneal neovascular growth [15, 28].

7. The corneal pocket assay can be used for safety assessment of synthetic or natural biomaterials in the form of injectable gel or solid material implantable in the micropocket. Induction of angiogenesis and inflammatory features can give info on biomaterial biocompatibility.

8. The effect of local drug treatment on corneal neovascularization may be studied in the form of ocular drops or ointments [22] or microinjection in the corneal thickness [23]. The effect of systemic drug treatment on corneal angiogenesis may be also evaluated [15, 17, 19]. However, when considering the size of the animals, systemic drug treatment in rabbits requires a higher amount of drugs than that for smaller animals, thus more expensive.

9. The corneal angiogenesis assay requires high technical competences. To run the rabbit cornea assay, researchers have to consider various critical requirements: animal facility equipped to maintain rabbits, technicians and researchers trained to manipulate rabbits, sterile surgery room and equipment, slit lamp, and video imaging system. Moreover, a project design should consider the costs for animal purchase, their maintenance and disposal, and drugs or substances to be used for surgery and experimental treatment. Beside this, as stated in Subheading 2, all the procedures need the approval by the local ethical board and the national agency or Ministry which acknowledges the European or other international guidelines and regulation for animal welfare, and in some countries (as Italy) the rules are very stringent.

4   Notes

1. Cornea has been found avascular in all strains of rabbits examined so far. In albino rabbits the newly formed vessels are clearly visible on the background of the iris.

2. Operator skill for pellet manipulation, surgery, and monitoring of angiogenesis is required. All procedures and observations are conducted in a double-masked manner and the code identities are revealed only after the end of the experiment and data elaboration.

3. Body weight: in the range 1.8–2.5 kg for an easy handling and prompt recovery from anesthesia. Sex: except when hormone dependency of cells or tumors is a prerequisite of the experimental setting, males are used. Check with your animal facility and veterinary doctor whether only SPF animals are admitted.

4. Sterility of materials and procedures is crucial to avoid nonspecific responses. DMSO should be avoided since incompatible with Elvax 40 polymerization and handling.

5. Ethylene vinyl acetate (Elvax 40) is the copolymer of ethylene and vinyl acetate (40%). Elvax 40 is a polymer that approaches elastomeric materials in softness and flexibility, used in biomedical engineering applications as a drug slow delivery device. While the polymer is not biodegradable within the body, it is quite inert and causes little or no reaction following implantation.

   Polyvinyl alcohol and polyhydroxyethyl-methacrylate (Hydron) can be used instead of Elvax 40 [29]. Hydron is hydrophobic; however, when the polymer is subjected to water it will swell due to the molecule's hydrophilic pendant group. Depending on the physical and chemical structure of the polymer, it is capable of absorbing from 10 to 600% water relative to the dry weight. Because of this property, it was one of the first materials to be successfully used in the manufacture of flexible contact lenses. When comparing the release kinetics of proteins from Elvax 40 and Hydron, the release from Hydron was the most rapid, while it was the slowest from Elvax, which continued to release the incorporated protein till 100 days [8]. Polyvinyl alcohol had an intermediate behavior [8].

In our experience, the polymer of hydroxyethyl-methacrylate gave less satisfactory results than Elvax 40. The release of molecules (i.e., gangliosides) from Elvax 40 pellets implanted in the cornea is in the order of 30–40% in the first 48 h and then it remains constant [30]. Initially the most superficial molecules are released, then water moves from the tissue inside the pellet matrix, leading the inner molecules on pellet surface [8].

6. Elvax 40 preparation and testing:
   - Weight 1 g of Elvax 40 (purchased as dry beads), extensively wash it in absolute alcohol for 100-fold at 37 °C, and dissolve in 10 ml of methylene chloride to prepare a 10% casting stock solution.
   - Test the Elvax 40 preparation for its biocompatibility [8]. The casting solution is eligible for routine use if no implant performed with this preparation induces the slightest or histological reaction in the rabbit cornea during 14 day examination period.

7. Variability among growth factors in inducing angiogenesis has been found considering different angiogenic factors, different providers and batch of preparation. Usually the dose of VEGF or FGF-2 able to give a positive angiogenic response varies in the range 200–400 ng/pellet [17].

8. When two factors are coreleased from the same pellet, the advice is to check before pellet implant if drug release in vitro is modified respect to the single molecule, as described in [25].

9. Immobilization (in appropriate contention box) during anesthetic procedure and observation is important to avoid self-induced injury.

10. Make the cut in the cornea in correspondence of the pupil and orient the micropocket toward the lower eyelid for an easy daily observation. It is important to calibrate the appropriate depth of the incision to avoid eye rupture.

11. The modulation of the angiogenic responses by different stimuli can be assessed in the rabbit cornea assay (a) by implanting single pellets releasing both the angiogenic stimulus and

the inhibitor [31–33], (b) by implanting in the same cornea two pellets placed in parallel micropockets and releasing different molecules [30, 34], and (c) through the addition or removal of single pellets in multiple implants [30].

12. Before embedding in O.C.T. Tissue-Tek medium, pellets should be removed and corneas sampled and marked (i.e., with a cotton thread) for subsequent orientation at the cryostat once embedded in O.C.T. medium.

13. Validated antibodies in our experience are anti-CD31 Ab (Dako, 200 µg/ml) (marker of neovascularization), anti-RAM11 Ab (Dako, 1.2 µg/ml) (marker of inflammation) (anti-α5β1 integrin Ab, Chemicon, 1:50) (adhesion molecule expressed in epithelial and endothelial cells).

14. When using eye ointment, take into account that paraffin-based preparations have to be used for short time (1 week) to avoid toxicity by excipients.

Acknowledgments

The work was supported by the Italian Ministry of University (MIUR) and the Italian Association for Cancer Research (AIRC). We thank Dr. Dario Rusciano, Sooft Italia SpA, for providing UPARANT.

References

1. Jain RK, Schlenger K, Hockel M et al (1997) Quantitative angiogenesis assays: progress and problems. Nat Med 3:1203–1208

2. Nowak-Sliwinska P, Alitalo K, Allen E et al (2018) Consensus guidelines for the use and interpretation of angiogenesis assays. Angiogenesis 21(3):425–532. https://doi.org/10.1007/s10456-018-9613-x

3. Chang JH, Gabison EE, Kato T et al (2001) Corneal neovascularization. Curr Opin Ophthalmol 12:242–249

4. Ellenberg D, Azar DT, Hallak JA et al (2010) Novel aspects of corneal angiogenic and lymphangiogenic privilege. Prog Retin Eye Res 29(3):208–248

5. Maddula S, Davis DK, Maddula S et al (2011) Horizons in therapy for corneal angiogenesis. Ophthalmology 118:591–599

6. Ambati BK, Nozaki M, Singh N et al (2006) Corneal avascularity is due to soluble VEGF receptor-1. Nature 443(7114):993–997

7. Gimbrone M Jr, Cotran R, Leapman SB et al (1974) Tumor growth and neovascularization: an experimental model using the rabbit cornea. J Natl Cancer Inst 52:413–427

8. Langer R, Folkman J (1976) Polymers for the sustained release of proteins and other macromolecules. Nature 363:797–800

9. Brem SS, Gullino PM, Medina D (1977) Angiogenesis: a marker for neoplastic transformation of mammary papillary hyperplasia. Science 195(4281):880–882

10. Cervenak L, Morbidelli L, Donati D et al (2000) Abolished angiogenicity and tumorigenicity of Burkitt lymphoma by Interleukin-10. Blood 96:2568–2573

11. Woolard J, Wang WY, Bevan HS et al (2004) VEGF165b, an inhibitory vascular endothelial growth factor splice variant: mechanism of action, in vivo effect on angiogenesis and endogenous protein expression. Cancer Res 64(21):7822–7835

12. Marconcini L, Marchio S, Morbidelli L et al (1999) c-fos-induced growth factor/vascular endothelial growth factor D induces angiogenesis in vivo and in vitro. Proc Natl Acad Sci U S A 96(17):9671–9676

13. Brem H, Folkman J (1975) Inhibition of tumor angiogenesis mediated by cartilage. J Exp Med 141(2):427–439

14. Bard RH, Mydlo JH, Freed SZ (1986) Detection of tumor angiogenesis factor in adenocarcinoma of kidney. Urology 27(5):447–450

15. Gallo O, Masini E, Morbidelli L et al (1998) Role of nitric oxide in angiogenesis and tumor progression in head and neck cancer. J Natl Cancer Inst 90:587–596

16. da Silva BB, da Silva Júnior RG, Borges US et al (2005) Quantification of angiogenesis induced in rabbit cornea by breast carcinoma of women treated with tamoxifen. J Surg Oncol 90(2):77–80

17. Ziche M, Morbidelli L, Choudhuri R et al (1997) Nitric oxide-synthase lies downstream of vascular endothelial growth factor but not basic fibroblast growth factor induced angiogenesis. J Clin Invest 99:2625–2634

18. Lasagna N, Fantappiè O, Solazzo M et al (2006) Hepatocyte growth factor and inducible nitric oxide synthase are involved in multidrug resistance-induced angiogenesis in hepatocellular carcinoma cell lines. Cancer Res 66(5):2673–2682

19. Ziche M, Morbidelli L, Masini E et al (1994) Nitric oxide mediates angiogenesis in vivo and endothelial cell growth and migration in vitro promoted by substance P. J Clin Invest 94:2036–2044

20. Monti M, Donnini S, Morbidelli L et al (2013) PKCε activation promotes FGF-2 exocytosis and induces endothelial cell proliferation and sprouting. J Mol Cell Cardiol 63:107–117

21. Carriero MV, Bifulco K, Minopoli M et al (2014) UPARANT: a urokinase receptor-derived peptide inhibitor of VEGF-driven angiogenesis with enhanced stability and in vitro and in vivo potency. Mol Cancer Ther 13(5):1092–1104. https://doi.org/10.1158/1535-7163.MCT-13-0949

22. Presta M, Rusnati M, Belleri M et al (1999) Purine analog 6-methylmercaptopurine ribose inhibits early and late phases of the angiogenesis process. Cancer Res 59(10):2417–2424

23. Ziche M, Morbidelli L (2009) Molecular regulation of tumour angiogenesis by nitric oxide. Eur Cytokine Netw 20(4):164–170

24. Chang LK, Garcia-Cardena G, Farnebo F et al (2004) Dose-dependent response of FGF-2 for lymphangiogenesis. Proc Natl Acad Sci U S A 101(32):11658–11663. https://doi.org/10.1073/pnas.0404272101

25. Taraboletti G, Morbidelli L, Donnini S et al (2000) The heparin binding 25kDa fragment of thrombospondin-1 promotes angiogenesis and modulates gelatinases and TIMP-2 in endothelial cells. FASEB J 14:1674–1676

26. Ziche M, Jones J, Gullino PM (1982) Role of prostaglandinE1 and copper in angiogenesis. J Natl Cancer Inst 69:475–482

27. Parenti A, Morbidelli L, Ledda F et al (2001) The bradykinin/B1 receptor promotes angiogenesis by upregulation of endogenous FGF-2 in endothelium via the nitric oxide synthase pathway. FASEB J 15(8):1487–1489

28. Federman JL, Brown GC, Felberg NT et al (1980) Experimental ocular angiogenesis. Am J Ophthalmol 89(2):231–237

29. Rogers MS, Birsner AE, D'Amato RJ (2007) The mouse cornea micropocket angiogenesis assay. Nat Protoc 2(10):2545–2550

30. Ziche M, Alessandri G, Gullino PM (1989) Gangliosides promote the angiogenic response. Lab Investig 61:629–634

31. Morbidelli L, Donnini S, Chillemi F et al (2003) Angiosuppressive and angiostimulatory effects exerted by synthetic partial sequences of endostatin. Clin Cancer Res 9(14):5358–5369

32. Bagli E, Stefaniotou M, Morbidelli L et al (2004) Luteolin inhibits vascular endothelial growth factor-induced angiogenesis; inhibition of endothelial cell survival and proliferation by targeting phosphatidylinositol 3′-kinase activity. Cancer Res 64(21):7936–7946

33. Donnini S, Finetti F, Lusini L et al (2006) Divergent effects of quercetin conjugates on angiogenesis. Br J Nutr 95(5):1016–1023

34. Cantara S, Donnini S, Morbidelli L et al (2004) Physiological levels of amyloid peptides stimulate the angiogenic response through FGF-2. FASEB J 18(15):1943–1945

Chapter 9

Avians as a Model System of Vascular Development

Rieko Asai, Michael Bressan, and Takashi Mikawa

Abstract

For more than 2000 years, the avian embryo has helped scientists understand questions of developmental and cell biology. As early as 350 BC Aristotle described embryonic development inside a chicken egg (Aristotle, Generation of animals. Loeb Classical Library (translated), vol. 8, 1943). In the seventeenth century, Marcello Malpighi, referred to as the father of embryology, first diagramed the microscopic morphogenesis of the chick embryo, including extensive characterization of the cardiovascular system (Pearce Eur Neurol 58(4):253–255, 2007; West, Am J Physiol Lung Cell Mol Physiol 304(6): L383–L390, 2016). The ease of accessibility to the embryo and similarity to mammalian development have made avians a powerful system among model organisms. Currently, a unique combination of classical and modern techniques is employed for investigation of the vascular system in the avian embryo. Here, we will introduce the essential techniques of embryonic manipulation for experimental study in vascular biology.

Key words Vascular development, Avian system, Cell labeling, Somatic transgenesis, Electroporation, Transduction

1 Introduction

Of the numerous animal models, the avian embryo represents an excellent system to study vascular biology particularly in the context of developmental biology [1–3]. The studies performed in birds have led to many of fundamental discoveries regarding the embryonic origins of the cardiovascular system (e.g., cardiomyocytes, blood, and endothelial cells) [4–8]. The advantages of the avian system (e.g., chick and quail), are related to the unique accessibility of the vascularized tissues in the embryo and the extra embryonic membranes, meaning that the cardiovascular system can be directly viewed by simply opening the eggshell. In addition to the ease of manipulation, the avian embryo is robust enough to continuously develop even after severe microsurgeries. In terms of acquisition, eggs can be consistently obtained from local farms year-round and be stored for up to 7 days before use. Once the eggs are transferred and incubated at 38 °C, the

Domenico Ribatti (ed.), *Vascular Morphogenesis: Methods and Protocols*, Methods in Molecular Biology, vol. 2206, https://doi.org/10.1007/978-1-0716-0916-3_9, © Springer Science+Business Media, LLC, part of Springer Nature 2021

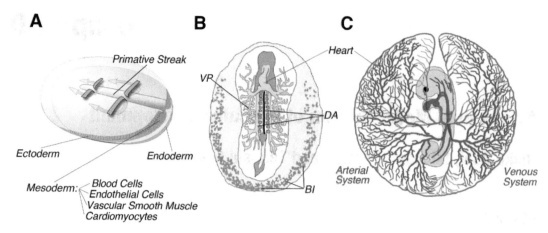

Fig. 1 Early cardiovascular development in avians. (**a**) Diagram of avian gastrulation. Epiblast cells ingress through the primitive streak generating the three germ layers (ectoderm, mesoderm, and endoderm). Cardiovascular lineages including blood, endothelium, smooth muscle, and cardiac muscle are derived from the mesoderm. (**b**) Shortly after gastrulation vasculogenesis initiates. The first major blood vessels, the paired dorsal aorta (DA) form adjacent to the embryonic midline. These vessels are flanked by a dense vascular plexus (VP). Coincident with these events the extraembryonic blood islands (BI), which give rise to primitive blood and endothelium, form along the posterior periphery of the embryo. (**c**) As the embryo increases in size the extraembryonic vascular bed extends out from the embryonic disk forming and arterial and venous system for gas and waste exchange

development progresses to a desired stage. The avian embryo, which is a flattened and two-layered blastoderm at the time of laid, develops rapidly to form the cardiovascular system following approximately 24–48 h of incubation (Fig. 1) [9–12]. Within the first 24–30 h of incubation, blood and vascular development initiates both in the form of extraembryonic blood islands and intraembryonic endothelial cell specification and differentiation. The first major blood vessels of the embryo, the paired dorsal aorta, have formed by 30–35 h of development, and a beating heart is apparent by approximately 40 h of development. The embryo is semitransparency throughout the early developmental processes, allowing for direct imaging intraembryonic tissues using a wide variety of microscopic techniques. Furthermore, modern advances including the creation of the annotated chick genome [13], gene expression profiling, live imaging, improved somatic trans genesis, and gene-specific attenuation have greatly expanded the types of experiments that can now be conducted in avians. Given that vascular patterning is highly conserved among amniotes [14], the value of the avian embryo is particularly relevant to conduct studies that would be challenging in mammals.

2 Materials

2.1 Fertilized Eggs

Freshly laid eggs of domestic chicken (*Gallus gallus*) and Japanese quail (*Coturnix japonica*) can be obtained from local farms. Specific pathogen free (SPF) chicken eggs are also commercially available from vendors including Charles River. Delivered eggs can be stores at 12–16 °C for up to a week. Once the eggs are transferred to a humidified incubator at 38 °C, they reinitiate their development. Embryos are typically staged based on the criteria established by Hamburger and Hamilton staging [10].

2.2 Incubators

A variety of incubators are available for avian development. Low temperature incubators, biomedical refrigerators, which can be set at 12–16 °C, even decided wine coolers, work for the storage of eggs. Wine coolers, which are not as expensive as standard biomedical grade refrigerators can also be used. For egg incubation, incubators (e.g., MIR-254, PHC, Fig. 2a) should be maintained at 38 ± 1 °C with humidity as created by placing trays of water in the bottom of the space. This is essential for preventing the drying and death of the embryos. Smaller format Styrofoam incubators (e.g., Thermal Air Incubator 1602N, Hova Bator, Fig. 2b) are also useful. When plugged into traceable outlet controllers (e.g., digital programmable timer, TM01715D, Hydroform, Fig. 2c), these small incubators allow for the precise set point /timing of incubation. Automatic egg rocking platforms are required, but preferred for incubation into later stage development (post embryonic day 10).

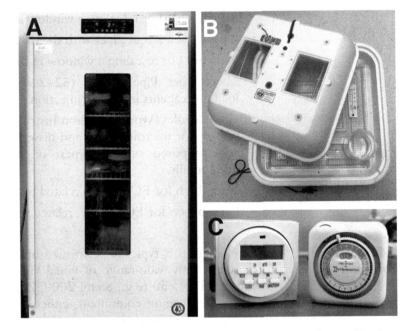

Fig. 2 Incubators. (**a**) Larger format incubator. (**b**) Smaller format Styrofoam incubator. (**c**) Weekly digital timer (left) and Daily timer (right)

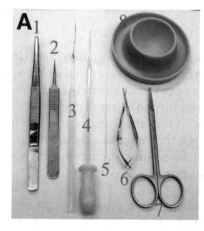

1. Blunt forceps

2. Fine forceps

3. Tungsten needle

4. Pasteur pipette

5. Glass needle

6. Spring scissors

7. Fine scissors

8. Egg holder

Fig. 3 Dissection tools. (**a**) Dissection tools. (**b**) Needle puller to make glass needles for injection (vertical type)

2.3 Dissection Tools (Fig. 3)

- Forceps: In general, a relatively blunt pair (for cracking the eggshell) and a fine pair (for microdissection, e.g., Dumont #5) are recommended.

- Scissors: Spring-balanced scissors with a blade length of 2–3 mm (e.g., Fine Scientific Instruments, 15000-08) work well for general dissection. Curved 4″ iris scissors (e.g., Fine Scientific Instruments, 14090-09) are used for cutting the eggshell when making a small access window.

- Tungsten needles: For dissecting tissues and cutting the vitelline membrane.

- Egg holder (metal or silicone).

- Scotch tape: For resealing a window in the eggshell.

- Disposable Pasteur pipet with the tip fire polished.

- Parafilm: For resealing a window in the eggshell.

- Microloader Pipette Tips (5242-956-003, Eppendorf): For loading reagents into glass injection needles.

- Glass needles (World Precision Instruments) and a micropipette puller: For microinjection and dissection. Micropipette pullers are composed of two types: oriented either vertically or horizontally.

- Water bath for EC culture, related to Subheading 3.1.3.

- Filter paper: for EC culture, related to Subheading 3.1.3.

2.4 Optical Equipment

Manipulation is typically performed under a dissecting stereomicroscope with wide-range of visual viewing magnification, from about ×6 to ×50 (e.g., Stemi 2000, Zeiss, MZ6, Leica). Attachments for imaging equipment, either for photo or video is very useful. Illumination is best obtained from a cool light source with

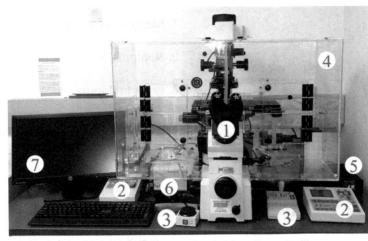

1. **Inverted Microscope**

2. **Controllers for objectives and shutter**

3. **Stage controller**

4. **Microscope incubator**

5. **Controller for heating system**

6. **Camera**

7. **Monitor**

Fig. 4 Equipment for live imaging. Example of imaging system for live imaging of ex ovo cultured embryos (related to Subheading 3.1.3)

dual gooseneck optic fibers. A blue light filter can be useful for detecting embryonic structures at early stage, such as a primitive streak.

For live imaging, images are captured with an inverted fluorescence microscope (e.g., Nikon, Ti, and Eclipse TE2000-E inverted fluorescence Microscope) linked to a high sensitivity camera (e.g., ANDOR iXon camera and Hamamatsu ORCA-Flash 2.8). The optical system is enclosed in a heating chamber (Fig. 4) [15]. Data from the imaging experiments are analyzed by various software, such as Imaris, Matlab, and ImageJ.

2.5 Electroporation/ Viral Injection Equipment (Fig. 5)

- Dissecting stereomicroscope: Leica MZ6 (Leica Microsystems).
- Electroporator: Nepa21 (Nepagene).
- Electrodes: made from platinum wire (0.25 or 0.5 mm diameter).
- Glass needles (World Precision Instruments).
- Pressure Injector: Eppendorf FemtoJet (Eppendorf).
- Manipulators: Leica micromanipulator (Leica Microsystems).
- Remote injector: Eppendorf Transmaster NK (Eppendorf).

2.6 Solutions

All of solutions should be sterilized.

1. 70% ethanol: For sanitizing the surface of eggs.

2. Tyrode's solution (10× stock): For in ovo manipulation, embryo isolation, or electroporation.

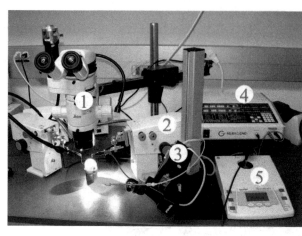

1. Stereo Microscope

2. Manipulator

3. Remote Controlled
   Micro-manipulator

4. Electroporator

5. Controller for
   Micro-manipulator

Fig. 5 Equipment for viral microinjection and electroporation. Example of a work station for viral microinjection and plasmid electroporation

80 g	NaCl
2.71 g	KCl
2.71 g	$CaCl_2 \cdot 2H_2O$
0.5 g	$NaH_2PO_4 \cdot 2H_2O$
2 g	$MgCl_2 \cdot 6H_2O$
10 g	Glucose
Distilled water to 1 l	

3. Dilute this stock with ten times volume of sterile distilled water prior to use. For preventing contamination, sterile both the 10× stock and ready-to-use buffer by filtration and keep them at 4 °C.

4. Pannett–Compton's saline: For isolation and culture of early avian blastoderm.

Stock solution A:	
121 g	NaCl
15.5 g	KCl
10.42 g	$CaCl_2 \cdot 2H_2O$
12.7 g	$MgCl_2 \cdot 6H_2O$
Distilled water to 1 l	
Stock solution B:	
2.365 g	$Na_2HPO_4 \cdot 2H_2O$
0.188 g	$NaH_2PO_4 \cdot 2H_2O$
Distilled water to 1 l	

Prior to use, mix 80 ml of stock solution A, B, and distilled water in the ratio of 4:6:90.

5. 1–10% India fountain ink in saline: For injecting into the yolk below the embryo in order to visualize tissues during in ovo manipulation (*see* Subheading 3.1.1). India ink can be found at art supply stores; avoid the waterproof varieties. Brands should be testing for toxicity by injecting into the yolks, incubating further and analyzing the resulting embryos.

6. 0.1–1% Fast green (Sigma): For Injecting into the yolk below the embryo to allow tissue visualization (it can be used in place of India ink) and for adding color to the DNA solution during electroporation.

7. 50% sucrose: For making the electroporation DNA solution.

2.7 Preparation for Agar–Albumen Culture Dishes for Modified New culture (EC Culture, Related to Subheading 3.1.3)

Protocol for making EC culture plates is following Subheading 3.1.3.

35 mm petri dishes (Falcon, 3001).

50 ml Falcon tubes (Falcon, 2098).

50 ml thin albumen, collected from about five unincubated eggs.

50 ml distilled water.

6 ml 1 M NaCl (dilute 5 M NaCl).

3 ml 10% glucose in H_2O (1 g/10 ml of H_2O).

0.3 g agarose (or Bacto agar, 214010, BD).

2.8 Reagents of Viral Production in Replication-Incompetent Retroviral System Using (Related to Subheading 3.3.4)

Protocol for making virus is following Subheading 3.3.4.
Plasmids

1. pSNID: SNV-gfp-MCS (Addgene, #53122), SNV-memTdTomato (Addgene, #53123), SNV-memTdTomato-2AcytoGFP (Addgene, #53124), SNV-lacZ (Addgene, #53125), more information about the vector in [16].

2. VSV-G: pCI-VSVG (Addgene, #1733).

Cell lines

3. Packaging cell lines: Phoenix-GP (ATCC, CRL-3215) and Phoenix-AMPHO (ATCC, CRL-3213).

4. Titering cell line: D17 (ATCC, CCL183).

Reagents

5. PEI Linear polyethylenimine MW 25000, transfection grade (Polyscience, 23966): Make up 1 mg/ml solution in water, neutralize with HCl and filter sterilize. It is stable at 4 °C at least 3 months, bulk of it stored at −80 °C.

6. Polybrene (Hexadimethrine bromide, Sigma H9268).

3   Techniques for Handling the Embryo

Avian embryology has a long history, and various manipulation techniques have been well established by many investigators. Below we provide a brief explanation of in ovo/ex ovo culture systems, basic experimental techniques for handling chick embryos, and somatic transgenesis and genomics in chick. These methods are also applicable to the quail embryo with some modifications.

3.1   Culture of the Avian Embryo

A variety of experimental manipulations are well tolerated in the developing avian embryo. While dozens of protocols exist for basic manipulation, most are variations of three basic systems: in ovo manipulation, shell-less culture of the embryo, and New culture (Fig. 6). The choice of which system should be used depends on the question being addressed and the developmental window during which the manipulations need to be made. As such, these features must be evaluated by the researcher prior to determining the proper technique to utilize. Below the protocols will be shown and the strengths and limitations of each system will be briefly discussed.

3.1.1   In Ovo Culture: Preparation, In Ovo Operation, and Incubation After the Surgery

As described above, embryos can be brought to a desired developmental stage with a high degree of reproducibility based on the timing of incubation. Once the stage at which an investigator wishes to manipulate the embryo is reached, the egg can be removed from the incubator and "windowed" for subsequent experiments (Fig. 6a). The hole in the eggshell can be resealed, and the embryo can be placed back in the incubator to develop further (Fig. 7).

1. Fertilized chicken eggs are incubated with their long axis horizontally in a humidified incubator at 38 °C (*see* **Note 1**).

2. After incubation for the desired length of time, place the egg in its side in an egg holder and wipe with 70% ethanol.

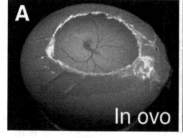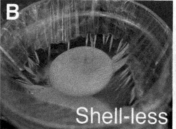

Fig. 6 Techniques for embryonic manipulation. (**a**) A windowed egg prepared for in ovo manipulation. (**b**) Embryo maintained in shell-less culture. (**c**) Modified new culture (EC culture)

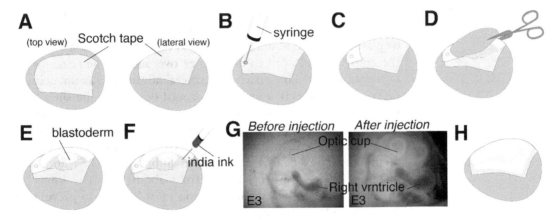

Fig. 7 Preparation of embryo for in ovo manipulation and subsequent culture. (**a**) After cleaning the surface of an eggshell, attach scotch tape on the top side. (**b**) Make a small hole and drain the thin albumin (approximately 3 ml). (**c**) Reseal the hole with a piece of scotch tape. (**d**) Cut out a circle from the tape-coated shell. (**e**) Place the blastoderm right under the window. (**f**) Inject India ink into the yolk below the blastoderm. (**g**) Visualizing the chick embryo of embryonic day 3 (E3) by injecting India ink. (**h**) After manipulation, reseal the hole and incubate in a humidified incubator at 38 °C

3. Attach scotch tape tightly to the surface of the shell in the upper side (Fig. 7a) (*see* **Note 2**).

4. Make a small hole (about 1 mm) on the shell in the pointed pole of the egg and gently drain about 1–3 ml of thin albumen, using a 5 ml syringe with a 18 G ×1–1/2 hypodermic needle (precisionglide needles, 305196, BD). Close the hole with a piece of scotch tape (Fig. 7b) (*see* **Note 3**).

5. Cut out a circle of about 2 cm diameter from the tape-coated shell. Rotate the egg until the blastoderm area is positioned right under the window (Fig. 7c–e).

6. Keep the embryo moist by periodically, adding a few drops of saline onto it.

7. Inject diluted India ink beneath the blastoderm gently, using a 1 ml syringe with a 31 G needle to visualize the embryo clearly (Fig. 7f, g).

8. The eggs can then be set under the optical equipment for surgeries.

9. After surgery, gently add a few drops saline and reseal the window with scotch tape or Parafilm carefully and tightly. Incubate the eggs in a humidified incubator at 38 °C (Fig. 7h) (*see* **Note 4**).

Whereas in ovo manipulations have the advantage of high viability in long-term post-manipulation incubation, it is often difficult to perform high-resolution microscopy on embryos in ovo. This problem is solved using the ex ovo culture methods shown in Subheadings 3.1.2 and 3.1.3.

3.1.2 Shell-Less Culture   In the shell-less culture techniques, the embryo, yolk, and albumin are transferred to an artificial container, such as a petri dish [17, 18] or plastic wrap "hammock" (Fig. 6b) [19]. Once outside the shell, the embryo is allowed to develop ex ovo and can be reincubated until late stages. This system has several advantages as the accessibility of the embryo is greatly improved outside of the shell. The shell-less system allows the investigator to target a greater variety of tissues in the embryo and is particularly useful for techniques that are related to the extraembryonic vasculature of the chorioallantoic membrane assay (CAM assay, *see* in Subheading 3.5) and live-imaging [20]. However, long-term viability is often lower in shell-less cultures, and great attention must be paid to preventing the embryo from drying out.

1. Fertilized chick eggs are incubated to E2 or E3 in a humidified incubated at 38 °C.

2. After incubation, sterilize the surface of the eggshell with 70% ethanol.

3. Crack the eggshell and slowly transfer the whole contents onto a petri dish (100 mm diameter and 20 mm high), making sure to keep the egg in the same orientation as the embryo has to be on the top of the yolk during the preparation of the culture.

4. Put on a lid of the petri dish and place the culture dish in a humidified incubator at 38 °C. Note: to decrease the chances of contamination, a few hundred microliters of PBS containing penicillin/streptomycin (final concentration 1%) can be gently dropped onto the surface of the embryo before putting on the lid.

3.1.3 New Culture:
Modified EC Culture   In 1955 D.A.T. New developed a system for ex ovo culture of avian embryos in which the early blastoderm was removed from the yolk and cultured on a piece of vitelline membrane stretched across a grass ring [21]. This technique has been modified by several additional investigators for over half of a century. The following protocol, known as Modified EC culture, uses filter paper rings and albumen–agar plates (Fig. 6c) [22–24].

Making Culture Plates

1. Collect 50 ml of thin albumin from unincubated eggs into a conical tube (50 ml size) and keep the tube in a water bath at 50 °C.

2. Add 50 ml saline to a sterile 250 ml flask and heat it in a microwave (Be careful of overheating).

3. Add the agar and stir until completely dissolved.

4. Put the flask into the water bath for about 10 min to allow the liquid to equilibrate at 50 °C.

5. Transfer the albumin into the flask and mix together.

6. Aliquot the agar–albumen mixture into 35 mm sterile petri dishes on a flat surface.

7. Leave them for several hours or overnight at room temperature to dry then store at 4 °C. Plates are usable for a week.

Making Filter Paper Rings

1. Make a hole in the middle of the filter paper circle. (If the center is smaller, so that the embryo adheres tightly, or larger, so you have more room to see.)

2. Sterilize paper rings and store in a closed container.

Dissection of Embryos for EC Culture (Fig. 8)

1. Eggs are incubated with their long axis vertically (the blunt end is up) in a humidified incubator at 38 °C.

2. After incubation for the desired length of time, clean the surface of the eggshell with 70% ethanol.

3. Holding the eggs with its long axis oriented vertically (the blunt end is up), crack the top of shell, that is, the air chamber. Then, remove the thin and thick albumen from the egg (Fig. 8a–c).

4. Carefully wipe the rest of thick albumen from over the blastoderm with a Kimwipes (Fig. 8d).

5. Once the vitelline membrane on the blastoderm is cleaned, gently place a filter paper ring onto the site. If the vitelline membrane is too dry to attach the filter paper ring, drop a little saline on the ring after placing (Fig. 8e).

6. Cut the vitelline membranes around the filter paper (Fig. 8f).

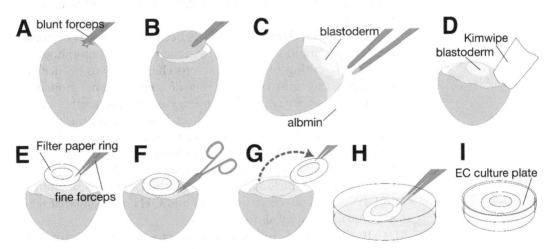

Fig. 8 Preparation of embryo for EC culture. (**a**) After cleaning the surface of an eggshell, crack the shell. (**b**) Cut out the top side of the shell. (**c**) Remove the thin and thick albumen. (**d**) Wipe the rest of thick albumen by Kimwipes. (**e**) Place a filter paper ring onto the blastoderm. (**f**) Cut the vitelline membranes around the filter paper. (**g**) Flip the filter paper upside down. (**h**) Gently wash away any attached yolk with a saline. (**i**) Place it on the agar–albumen in the EC culture dish with ventral side up and incubate in a humidified incubator at 38 °C

7. Gently pull the filter paper from the egg and flip the filter paper upside down (Fig. 8g).

8. Gently wash away any attached yolk with a saline (Fig. 8h).

9. Place the embryo and ring on the agar–albumen in the EC culture dish with ventral side up (Fig. 8i).

10. Keep the culture dish at 38 °C inside of a larger container with moistened tissues for added humidity.

This culture method is an extraordinarily powerful system as it allows for high-resolution examination of early developmental processes including cardiovascular development in situ. Again as the development occurs while the embryo is still largely a two-dimensional embryonic disk and the vast majority of yolk has been removed from the embryo, New culture is ideal for detailed embryonic observation. The major drawback of this system, however, is that embryos typically do not advance beyond the third day of development on the filter paper rings, and so it is not useful for studying later developmental processes.

3.2 Labeling Techniques

For labeling cells, traditionally either Nile Blue vital staining or sprinkled carbon particles on the surface of the chick embryo were used for tracing cell movement [25–27]. Currently, more sensitive fluorescent vital dies (e.g., DiI, DiO) have been developed and applied to cell labeling and tissue staining [28, 29]. Additionally, improved somatic transgenics, including viral transfection using replication-incompetent retroviral vectors and electroporation introducing reporter genes (e.g., GFP, LacZ), are also widely used for labeling cells (the methods are shown below in Subheading 3.4) [30, 31].

Quail-chick chimera system has been used in terms of examining cell migration and differentiation during morphogenesis [32, 33]. Grafting of quail cells into chick embryos or vice versa represents the most common way of chimera generation in the avians. The advantage of this strategy is that the grafted cell can be easily distinguished from the host tissues. Quail nuclei can be recognized by the presence of foci of heavily condensed heterochromatin that can be detected histologically [34]. As an improvement to identification, a variety of species-specific antibodies have been generated including the quail/chick perinuclear antibody (QCPN) which reacts to quail cells and QH1 antibody that recognizes quail endothelial cells [8, 35].

In this section, we introduce an easy and simple protocol of labeling cells using fluorescent vital dye.

3.2.1 Tracing Cells by Vital Dye Staining

1. Prepare embryos following Subheading 3.1.1.

2. To access the surface of the embryo, make a slit in the vitelline membrane and peel it off the area pellucida, using a tungsten

needle. Choose the embryos that will be used in the labeling and seal them with scotch tape to prevent drying, while the dyes are prepared.

3. Mix CM-DiI (Molecular Probes add catalog number) and DMSO (Wako) or Tetraglycol (Sigma, T3396)] to a final concentration of 1–2 mg/ml. Immediately before using, dilute this stock solution into saline (final 1–2 μM).

4. Load 5 μl of the mixture into a glass needle by a loading pipette tip.

5. Place the dye-filled glass needle into the pressure injector/micromanipulator and position it over the targeted region of the blastoderm (Fig. 5).

6. Adjust the z position until the needle is within the target tissue and inject.

7. Gently add a little saline onto the embryo for washing.

8. Seal the window with scotch tape again and incubate the eggs in a humidified incubator at 38 °C to a desired stage.

The labeling techniques using fluorescent tracers can be combined with current molecular biology techniques including live-imaging, IF, FACS, or RNA-seq to test cellular migration or identity during developmental processes.

3.3 Somatic Transgenesis

A transgenic approach has been essential for studying the function of genes and gene products in cells. Transfection is a process that introduces foreign nucleic acids into eukaryotic cells, and can be considered a temporary, or transient type of transgenesis. Methods of transfection are generally classified into three groups: physical, chemical and biological [36, 37]. Physical transfection methods include electroporation, microinjection and biolistic particle delivery. These methods require special equipment. Chemical transfection methods use cationic polymer, cationic lipid and calcium phosphate. The underlying principle of these methods is similar. Chemicals having a positive charge make complexes with negatively charged nucleic acids and these are taken into cells by phagocytosis and endocytosis. Biological transfection methods use a viral-mediated gene delivery system, known as transduction, and are commonly applied in clinical research, including for gene therapy [38]. Transfection is highly efficient and applicable to many cell types both in vivo and in vitro.

In the chick embryo, several efficient methods have been developed for timing- and spatial-specificity [16, 39–43]. Electroporation, lipofection, and adenoviral-mediated gene transfer are effective gene delivery systems which can be applied to both dividing- and nondividing cells. In these methods, transferred genes are not incorporated into the host genome. This means that the transgene expression is quick and suited for short-term analysis; however, the expression is transient. In contrast, retroviral-mediated gene transfer is more reliable for long-term analysis. The retroviral RNA genome is reverse transcribed into a double-stranded DNA which is integrated into the host chromosome, leading to stable misexpression of the transgene in the host cells and its descendants. The entry of the retroviral DNA into the host nucleus requires a disassembly of the nuclear membrane associated with mitosis, except for lentivirus. Below, basic protocols are presented for each technique.

3.3.1 Electroporation

Electroporation uses high voltage and short electric pulses to transport the exogenous DNA into the cytoplasm of the cells [44]. While this technique allows a rapid gain-of-function and loss-of-function experiment using the efficient introduction of the exogenous genes, some limitations also exist. As described above, expression of the transgenes is transient; the expression usually peaks one day after electroporation [41] and gradually decreases, indicating degradation and dilution of the introduced gene. Additionally, the copy number of exogenous genes introduced into a cell can not be controlled precisely, leading to a variation in expression levels amongst electroplated cells.. Furthermore, some tissues are difficult to target, because of the embryonic topology and structure of the tissues. For example, highly efficient electroporation has been achieved in epithelial tissues surrounding a small lumen, such as a neural tube, in part because the injected DNA stays inside the cavity during electroporation. In contrast, injected DNA is diluted when injected onto epithelial sheets and electroporation is less effective. These problems may be overcome by modifying the electroporation system. Ex ovo electroporation and culture is also useful in terms of getting good accessibility to the embryo [45]. We show simplified protocols of both in ovo and ex ovo electroporation below.

Equipments and materials are shown in Subheading 2.5 (Figs. 5 and 9).

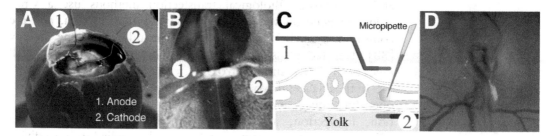

Fig. 9 In ovo electroporation. (**a**) Set up the electrodes in an egg. (**b**) Setting of the electrodes under a dissecting microscope. (**c**) Diagram of in ovo electroporation into embryo. The plasmid is injected into the embryonic coelom and current is applied across the embryo. The vitelline membrane on the embryo is cut before setting the electrodes. (**d**) GFP expression in the forelimb of a chick embryo after electroporation

In Ovo Electroporation
(Fig. 9)

This method is useful for an experiment that needs long term incubation after electroporation.

1. Prepare embryos following Subheading 3.1.1.

2. Set electric pulse (*see* **Note 5**). Mount the egg with an egg holder on the stage of a dissecting stereomicroscope. The equipment for electroporation is shown in Fig. 5.

3. Gently add a little saline on the embryo to prevent touching electrodes to the surface of the vitelline membrane directly.

4. Insert a cathode beneath the targeting site from the extra embryonic region; this will decrease tissue damage (Fig. 9a, b).

5. Place an anode over the targeting area in the embryo, placing it in a parallel position with the cathode. These electrodes must be in saline, not touching the embryo; this is to avoid tissue damage (Fig. 9a, b).

6. Inject the DNA mixture (final concentration: DNA 1–3 μg, 0.1% Fast green, 1% sucrose) to the targeting site in the embryo (Fig. 9c).

7. Send the electrical pulse. Bubbles will form on the electrodes.

8. After electroporation, seal the window in the shell with scotch tape and incubate in a humidified incubator at 38 °C.

Ex Ovo Electroporation
(Fig. 10)

This method is useful for experiments combined with high-resolution examination (e.g., live imaging) in early developmental processes [46].

1. Prepare the EC culture plate and embryo, attaching the filter paper ring, following Subheading 3.1.3.

2. Set electric pulse (*see* **Note 5**). Mount the EC culture plate on the stage of a dissecting stereomicroscope.

3. Insert a cathode in the middle of the agar plate (Fig. 10a).

4. Place the embryo on its filter paper ring on the agar in the EC culture plate. The targeting site is positioned over the cathode (Fig. 10a, b).

5. Carefully drip a little saline on the embryo.

6. Rearrange an anode over the targeting area in the embryo, placing it in a parallel position with the cathode. The electrode must be in saline, not touching the embryo.

7. Inject the DNA mixture to the targeting site in the embryo (Fig. 10c).

8. After electroporation, move the embryo to a new EC culture plate, and incubate in a humidified incubator at 38 °C.

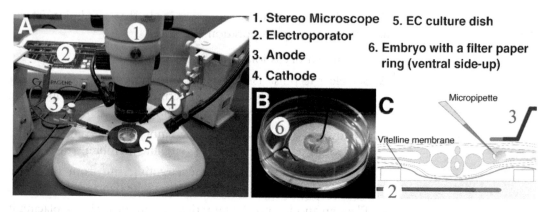

1. Stereo Microscope   5. EC culture dish
2. Electroporator
3. Anode   6. Embryo with a filter paper
4. Cathode   ring (ventral side-up)

Fig. 10 Ex ovo electroporation. (a) Basic equipment for ex ovo electroporation. (b) Setting the electrodes with an embryo in an EC culture dish. (c) Diagram of ex ovo electroporation into embryo. The plasmid is injected into the embryonic coelom and current is applied across the embryo

3.3.2 Lipofection

Lipofection is one of the most popular methods of transfection. The synthetic cationic lipids interact with negatively charged nucleic acids to form lipid–nucleic acid complexes [36, 37]. These complexes directly interact with the plasma membrane of the cell or are taken into by endocytosis. The method is simple; mix the reagents (e.g., Lipofectamine) and nucleic acids.

Chemical transfection including lipofection has been used in the chick embryo. However, these studies had relatively lower efficiency of transfection or the expression of the transferred gene was transient and lasting short period [42, 47, 48]. Current report shows that lipofection using a combination of Lipofectamine3000 and a pCAG expression vector achieved efficient transfection to cardiac tissues in the developing heart with no notable changing on the electrical activity of the heart [43]. Additionally, the piggy-Bac transposon vector system [49, 50] combined with mouse alpha-myosin Heavy Chain (aMHC) promoter [51, 52] led to genome integration of the transferred gene and cell-type specific targeting. These techniques would make transfection to the cardio-vascular tissues of chick embryo simpler, and allow gene editing technology such as TALEN and CRISPR/Cas9 to be easily applied to them.

3.3.3 Adenovirus

Adenoviruses carry a linear dsDNA genome of approximately 36 kb [38]. Engineered adenoviral vectors are designed to be replication incompetent by deletion of viral genes and replacement by transcription machinery to express transgenes of interest. Packaging cells, such as HEK293T, and the cotransfection of enveloping genes are required for replicating and packaging the viruses.

These viruses have broad cell type and tissue tropism, excellent gene-transfer efficiency in vivo and in vitro, and potential capacity for large foreign inserts. Virus titers as high as 10^7–10^8 infectious units (IU)/ml can be obtained, and increased ($>10^{10}$ IU/ml) by ultracentrifugation [39]. Moreover, adenoviral vector genomes reside in most infected cells as episomes, meaning that the copy number of episomes are diluted during cell proliferation, but the transgene expression begins quickly after the infection. Therefore, adenoviral-mediated gene transfer is a powerful tool for short-term studies, and has been used for overexpression in avians [53, 54].

3.3.4 Replication-Incompetent Retroviral-Mediated Gene Transfer

Retroviruses are enveloped, single-stranded RNA viruses, that are reverse transcribed into a double stranded DNA that has the ability to integrate into the host genome. Because of this strength, genetic manipulation in avians is classically performed using retroviral-mediated somatic transgenesis. As mentioned above, this feature allows stable expression of the transgenes in the infected cells.

Retroviral expression system consists of two types; replication-competent and replication-incompetent. Replication retroviruses, such as RCAS, can spread from initial infected cells to others, because of containing the critical genes required for viral propagation (*gag*, *pol*, and *env*). In contrast, replication-incompetent retroviruses have been engineered by lacking all or some of these viral genes, while retaining the cis elements required for viral packaging, transcription, and integration. Thus, replication-incompetent retroviruses enable only vertical transmission of the introduced gene, indicating that usually these viruses infect fewer cells (Fig. 11).

In this section, we outline a protocol for a recently established replication-incompetent retroviral system using pSNID and VSV-G (*see* in Subheading 2.8, Fig. 7) [16]. Usually, a high titer of replication-incompetent virus is required for getting efficient infection per injection, because only a few nanoliters of viral solution can be injected into a small avian embryo. The spleen necrosis virus (SNV) [55, 56] works well in avian tissues and the titer is higher than other popular replication-incompetent retroviruses [41]. The pSNID vector has tandem promoters derived from the Moloney Murine Leukemia Virus (MLV) and the SNV and the virus particles are pseudotyped with the vesicular stomatitis virus envelope protein (VGV-S). The pSNID/VSV-G system does not rely on a dedicated packaging cell line and the virus can be created in the HEK293 cell line. The pSNID tandem promoters show robust expression of the transgene in avian embryos in all tissues and does not appear to be silenced during development [16].

Transfection

This protocol is per one cell culture dish (100 mm), and four dishes are needed for making an appropriately high titer viral stock; therefore, multiply everything by 4. Information about materials are shown in Subheading 2.8 above.

1. Culture   Phoenix-GP   or   Phoenix-AMPHO   at   30–90% confluent.

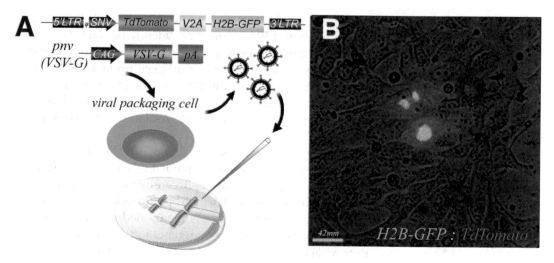

Fig. 11 Replication-incompetent retroviral-mediated gene transfer system. (**a**) Construction and injection of a replication-incompetent retrovirus. Viral plasmid lacking packaging genes is transfected into a packaging cell line. These cells produce viral particles which are collected and injected into an embryo where they infect and integrate into the host cell genome. (**b**) Example of cells infected with a viral vector containing H2b-GFP and a membrane-bound tdTomato

2. For preparing transfection, mix 4 μg of DNA, 100 μm of serum free DMEM, and 20–32 μl of PEI in a 1.5 ml eppendorf tube and vortex the tube.

3. Leave the tube at room temperature for 10–15 min.

4. Take off old media on the cell culture plates and add 10 ml of new media.

5. Add the PEI/DNA mixture to the cells, swirl and place back in the incubator.

6. After overnight culture, remove old media, and replace it with 5 ml of fresh media to each plate and place back in the incubator.

7. Next day, collect the media off the plates, and store at −80 °C, add 5 ml of fresh media to each plate and culture it overnight in the incubator.

8. Next day, collect the media off the plates, and store at −80 °C. You can save the plates to check the expression of the transfected plasmids.

Virus Production

To get sufficient titers, plan on concentrating the supernatant collected from 4 plates in the above steps by ultracentrifugation using an SW28 swing rotor.

1. Thaw the stored supernatant collected from four plates in the above steps, transfer it into an Ultra-Clear centrifuge tube (Beckman, 344058), and set with the SW28 tubes in a SW28

swinging-Bucket Rotor (Beckman, 342204). See the instructions of SW28 rotator system provided by Beckman for detailed steps of setting up and safety.

2. Spin for 2 h at 18,000 rpm (50,000 × g) at 4 °C (*see* **Note 6**).

3. Turn the tube right side up and let the media from the sides of the tube fall back down to the bottom. Expect between 40 and 60 μl. More DMEM can be added if desired.

4. Resuspend the pellet gently pipetting up and down without making bubbles. Measure volume and transfer to a microfuge tube.

5. Add the appropriate amount of Polybrene (Hexadimethrine bromide, Sigma, H9268) to a final concentration of 100 μg/ml, and store it at −80 °C. Note: To get higher titers for injection, do a first spin of the supernatant off, during that off, then add new supernatant off, and collect the combined pellet that results.

6. Check the titer of the virus using the on D17 dog epithelial cell line.

3.4 Genomics

Extremely efficient genomics and bioinformatics tools are currently available for a various model organisms, and avian embryology has successfully begun to adopt these techniques. For example, databases such as those hosted by the University of Arizona (Geisha) [57–59] provide powerful resources for the pattern of gene expression during embryogenesis. Furthermore, the advent of next-generation sequencing technologies such as RNAseq provides whole transcriptome quantification and profiling not previously available for studies in avians. Prior microarray-based large format expression profiling required the construction or use of preexisting chips with cDNA or oligonucleotide libraries spotted at specific positions. Due to poor availability of avian-specific chips, microarray studies were limited and cumbersome to all but a select few laboratories working in avian species. However, techniques such as RNAseq allow for massively parallel shotgun sequencing of total mRNA isolated from embryonic tissue. Bioinformatics is then used to determine the identity and relative levels of mRNA species within the sample. Thus, no species-specific cDNA library or chip is required. Accordingly, avians represent an exciting system for these global RNA sequencing techniques as their unparalleled accessibility as an amniote model system will uniquely allow for the detailed evaluation of expression changes across finely separated developmental stages, within closely opposed progenitor fields, and following experimental manipulation through simple isolation and comparison of embryonic tissues.

3.5 Chorioallantoic Membrane Assay (CAM)

During avian development the mesodermal layers of the allantois and chorion fuse to form the chorioallantoic membrane (CAM). This structure rapidly expands around the embryo, generating a rich vascular network that provides an interface for gas and waste exchange [60]. The CAM is an ideal in vivo system for studying vascular network formation as it is a thin, transparent, two-dimensional surface that is easily accessible (especially in shell-less culture). Furthermore, CAM development proceeds with the establishment of a functional immune system, meaning that foreign cells (including tumors cells) can be cultured directly on the CAM to analyze vascular invasion of these tissues [61–66]. As such, the capacity of the CAM vasculature to sprout or regress in response to ectopic cells and/or factors has been widely used as a vascular assay system. Using this system, the angiogenic or antiangiogenic potency of growth factors and pharmacological reagents can be quickly and reliably determined in a low-cost in vivo system [67]. Molecular signals that initiate tumor neovascularization and metastasis are also frequently studied using the CAM. Consequently, the CAM's accessibility and simplicity, in regard to experimental execution, provide an important in vivo functional assay through which the mechanisms of vascular sprouting can be examined.

3.6 Other Approaches

As discussed above, one of the principle strengths of the avian model is its amenability to physical manipulation. Ablation experiments, removing cells or tissues from the embryo, are often used to determine the capacity of one tissue to induce and/or suppress the fate decision in adjacent tissues [68, 69]. In transplantation, donor tissue from one embryo is removed and placed either into a host embryo or host region within the same embryo. The purposes of this experiment is various such as investigating tissue interaction and labeling cells combined with chick-quail chimera system or transgenic GFP chick [15, 29]. To study the function of the molecular signals, COS cells (derived from simian kidney tissue), which have had a gene of interest introduced [28, 70, 71], or microbeads soaked with signaling factors can be implanted into the tissues [72–74].

In many of the model systems in use, transgenic animals have been generated. Transgenic avians are also available [75–78] and allow us to investigate the mechanism of morphogenesis or gene functions [46, 79]. Gene editing tools such as CRISPR/Cas9 [80–82] and improved transfection methods will accelerate our ability to effectively develop transgenic avians and make the avian model system more valuable.

3.7 Conclusion

This chapter has provided some basic principles of avian experimental biology, with the hope of introducing readers to the utility, diversity, and limitations of the avian model system. This system is versatile, and classical embryology can be combined with modern molecular approaches to allow for the simultaneous investigation of multiple levels of biology (molecular through organismal). In many ways, it is this elegant blend of simplicity and complexity that has kept investigator returning to the avian embryo for the last 2000 years.

4 Notes

1. Embryos can develop even if the eggs are incubated with their long axis vertically; this can save space in the incubator.

2. The tape-coating can prevent fallen small shavings from the cutting shells, which can cause contamination.

3. Removing some of the albumin makes the embryo sink and leaves space for opening the window; the yolk exists right under the shell in the horizontally laid eggs.

4. Parafilm creates a tight seal on eggs. It cannot be sterilized, but if the paper-facing side of the Parafilm is used, there is usually very little contamination. Add a second layer if you find that the Parafilm splits in the heat of the incubator.

5. Pulse setting in electroporator depends on tissues and developmental stage.

6. The viral pellet is very delicate. Be at the machine right when the spin ends. Once the spin ends, remove tubes, immediately pour off the supernatant and let the tube sit upside down on a Kimwipes for one minute.

Acknowledgments

We thank J. Hyer, L. Hua, and Z. Zhao for their comments and past and present Mikawa lab members for their suggestions regarding this work, supported in part by Uehara Memorial Foundation Fellowship and JSPS Postdoctoral Fellowship for Research Abroad to R.A.; CDA34760248 from the American Heart Association, NIH-R00HL122360 and NIH-R01HL146626 to M.B.; and R37HL078921, R01HL112268, R01HL122375, R01HL132832, and R01HL148125 to T.M.

References

1. Aristotle PA (1943) Generation of animals. Loeb Classical Library (translated), vol 8. Harvard University Press, Cambridge, MA

2. Pearce JM (2007) Malpighi and the discovery of capillaries. Eur Neurol 58(4):253–255. https://doi.org/10.1159/000107974

3. West JB (2013) Marcello Malpighi and the discovery of the pulmonary capillaries and alveoli. Am J Physiol Lung Cell Mol Physiol 304(6):L383–L390. https://doi.org/10.1152/ajplung.00016.2013

4. Rawles ME (1943) The heart-forming areas of the early chick blastoderm. Physiol Zool 16 (1):22–43

5. Stalsberg H, DeHaan RL (1969) The precardiac areas and formation of the tubular heart in the chick embryo. Dev Biol 19(2):128–159. https://doi.org/10.1016/0012-1606(69)90052-9

6. Coffin JD, Poole TJ (1988) Embryonic vascular development: immunohistochemical identification of the origin and subsequent morphogenesis of the major vessel primordia in quail embryos. Development 102 (4):735–748

7. Noden DM (1989) Embryonic origins and assembly of blood vessels. Am Rev Respir Dis 140(4):1097–1103. https://doi.org/10.1164/ajrccm/140.4.1097

8. Pardanaud L, Yassine F, Dieterlen-Lievre F (1989) Relationship between vasculogenesis, angiogenesis and haemopoiesis during avian ontogeny. Development 105(3):473–485

9. Patten BM, Kramer TC (1933) The initiation of contraction in the embryonic chick heart. Am J Anat 53(3):349–375. https://doi.org/10.1002/aja.1000530302

10. Hamburger V, Hamilton HL (1992) A series of normal stages in the development of the chick embryo. 1951. Dev Dyn 195(4):231–272. https://doi.org/10.1002/aja.1001950404

11. Pardanaud L, Luton D, Prigent M, Bourcheix LM, Catala M, Dieterlen-Lievre F (1996) Two distinct endothelial lineages in ontogeny, one of them related to hemopoiesis. Development 122(5):1363–1371

12. Martinsen BJ (2005) Reference guide to the stages of chick heart embryology. Dev Dyn 233 (4):1217–1237. https://doi.org/10.1002/dvdy.20468

13. Wallis JW, Aerts J, Groenen MA, Crooijmans RP, Layman D, Graves TA, Scheer DE, Kremitzki C, Fedele MJ, Mudd NK, Cardenas M, Higginbotham J, Carter J, McGrane R, Gaige T, Mead K, Walker J, Albracht D, Davito J, Yang SP, Leong S, Chinwalla A, Sekhon M, Wylie K, Dodgson J, Romanov MN, Cheng H, de Jong PJ, Osoegawa K, Nefedov M, Zhang H, McPherson JD, Krzywinski M, Schein J, Hillier L, Mardis ER, Wilson RK, Warren WC (2004) A physical map of the chicken genome. Nature 432(7018):761–764. https://doi.org/10.1038/nature03030

14. Baldwin HS (1996) Early embryonic vascular development. Cardiovasc Res 31:E34–E45

15. Maya-Ramos L, Mikawa T (2020) Programmed cell death along the midline axis patterns ipsilaterality in gastrulation. Science 367(6474):197–200. https://doi.org/10.1126/science.aaw2731

16. Venters SJ, Dias da Silva MR, Hyer J (2008) Murine retroviruses re-engineered for lineage tracing and expression of toxic genes in the developing chick embryo. Dev Dyn 237 (11):3260–3269. https://doi.org/10.1002/dvdy.21766

17. Auerbach R, Kubai L, Knighton D, Folkman J (1974) A simple procedure for the long-term cultivation of chicken embryos. Dev Biol 41 (2):391–394. https://doi.org/10.1016/0012-1606(74)90316-9

18. Luo J, Redies C (2004) Overexpression of genes in Purkinje cells in the embryonic chicken cerebellum by in vivo electroporation. J Neurosci Methods 139(2):241–245. https://doi.org/10.1016/j.jneumeth.2004.04.032

19. Dugan JD Jr, Lawton MT, Glaser B, Brem H (1991) A new technique for explantation and in vitro cultivation of chicken embryos. Anat Rec 229(1):125–128. https://doi.org/10.1002/ar.1092290114

20. Sanders TA, Llagostera E, Barna M (2013) Specialized filopodia direct long-range transport of SHH during vertebrate tissue patterning. Nature 497(7451):628–632. https://doi.org/10.1038/nature12157

21. New DAT (1955) A new technique for the cultivation of the chick embryo *in vitro*. J Embryol Exp Morphol 3(4):326–331

22. Sundin O, Eichele G (1992) An early marker of axial pattern in the chick embryo and its respecification by retinoic acid. Development 114 (4):841–852

23. Chapman SC, Collignon J, Schoenwolf GC, Lumsden A (2001) Improved method for chick whole-embryo culture using a filter paper carrier. Dev Dyn 220(3):284–289. https://doi.org/10.1002/1097-0177(20010301)220:3<284::Aid-dvdy1102>3.0.Co;2-5

24. Kimura W, Yasugi S, Fukuda K (2007) Regional specification of the endoderm in the early chick embryo. Develop Growth Differ 49 (5):365–372. https://doi.org/10.1111/j.1440-169X.2007.00933.x

25. Gräper L (1929) Die Primitiventwicklung des Hühnchens nach stereokinematographischen Untersuchungen, kontrolliert durch vitale Farbmarkierung und verglichen mit der Entwicklung anderer Wirbeltiere. Wilhelm Roux Arch Entwickl Mech Org 116 (1):382–429. https://doi.org/10.1007/bf02145235

26. Wetzel R (1929) Untersuchungen am Hühnchen. Die Entwicklung des Keims während der ersten beiden Bruttage. Wilhelm Roux Arch Entwickl Mech Org 119 (1):188–321. https://doi.org/10.1007/bf02111186

27. Spratt NT Jr (1946) Formation of the primitive streak in the explanted chick blastoderm marked with carbon particles. J Exp Zool 103 (2):259–304. https://doi.org/10.1002/jez.1401030204

28. Bressan M, Liu G, Mikawa T (2013) Early mesodermal cues assign avian cardiac pacemaker fate potential in a tertiary heart field. Science 340(6133):744–748. https://doi.org/10.1126/science.1232877

29. Asai R, Haneda Y, Seya D, Arima Y, Fukuda K, Kurihara Y, Miyagawa-Tomita S, Kurihara H (2017) Amniogenic somatopleure: a novel origin of multiple cell lineages contributing to the cardiovascular system. Sci Rep 7(1):8955. https://doi.org/10.1038/s41598-017-08305-2

30. Mikawa T, Borisov A, Brown AM, Fischman DA (1992) Clonal analysis of cardiac morphogenesis in the chicken embryo using a replication-defective retrovirus: I. Formation of the ventricular myocardium. Dev Dyn 193 (1):11–23. https://doi.org/10.1002/aja.1001930104

31. Cohen-Gould L, Mikawa T (1996) The fate diversity of mesodermal cells within the heart field during chicken early embryogenesis. Dev Biol 177(1):265–273. https://doi.org/10.1006/dbio.1996.0161

32. Cox CM, Poole TJ (2000) Angioblast differentiation is influenced by the local environment: FGF-2 induces angioblasts and patterns vessel formation in the quail embryo. Dev Dyn 218 (2):371–382. https://doi.org/10.1002/(sici)1097-0177(200006)218:2<371::Aid-dvdy10>3.0.Co;2-z

33. Arima Y, Miyagawa-Tomita S, Maeda K, Asai R, Seya D, Minoux M, Rijli FM, Nishiyama K, Kim KS, Uchijima Y, Ogawa H, Kurihara Y, Kurihara H (2012) Preotic neural crest cells contribute to coronary artery smooth muscle involving endothelin signalling. Nat Commun 3:1267. https://doi.org/10.1038/ncomms2258

34. Le Douarin N (1973) A biological cell labeling technique and its use in experimental embryology. Dev Biol 30(1):217–222. https://doi.org/10.1016/0012-1606(73)90061-4

35. Peault BM, Thiery JP, Le Douarin NM (1983) Surface marker for hemopoietic and endothelial cell lineages in quail that is defined by a monoclonal antibody. Proc Natl Acad Sci U S A 80(10):2976–2980. https://doi.org/10.1073/pnas.80.10.2976

36. Hahn P, Scanlan E (2010) Gene delivery into mammalian cells: an overview on existing approaches employed in vitro and in vivo. In: Bielke W, Erbacher C (eds) Nucleic acid transfection. Springer, Berlin, pp 1–13. https://doi.org/10.1007/128_2010_71

37. Kim TK, Eberwine JH (2010) Mammalian cell transfection: the present and the future. Anal Bioanal Chem 397(8):3173–3178. https://doi.org/10.1007/s00216-010-3821-6

38. Lai CM, Lai YK, Rakoczy PE (2002) Adenovirus and adeno-associated virus vectors. DNA Cell Biol 21(12):895–913. https://doi.org/10.1089/104454902762053855

39. Yamagata M, Jaye DL, Sanes JR (1994) Gene transfer to avian embryos with a recombinant adenovirus. Dev Biol 166(1):355–359. https://doi.org/10.1006/dbio.1994.1321

40. Ogura T (2002) In vivo electroporation: a new frontier for gene delivery and embryology. Differentiation 70(4–5):163–171. https://doi.org/10.1046/j.1432-0436.2002.700406.x

41. Ishii Y, Mikawa T (2005) Somatic transgenesis in the avian model system. Birth Defects Res C Embryo Today 75(1):19–27. https://doi.org/10.1002/bdrc.20033

42. Ishii Y, Garriock RJ, Navetta AM, Coughlin LE, Mikawa T (2010) BMP signals promote proepicardial protrusion necessary for recruitment of coronary vessel and epicardial progenitors to the heart. Dev Cell 19(2):307–316. https://doi.org/10.1016/j.devcel.2010.07.017

43. Goudy J, Henley T, Mendez HG, Bressan M (2019) Simplified platform for mosaic in vivo analysis of cellular maturation in the developing heart. Sci Rep 9(1):10716. https://doi.org/10.1038/s41598-019-47009-7

44. Itasaki N, Bel-Vialar S, Krumlauf R (1999) 'Shocking' developments in chick embryology: electroporation and in ovo gene expression. Nat Cell Biol 1(8):E203–E207. https://doi.org/10.1038/70231

45. Fukuda K (2009) Electroporation of nucleic acids into chick endoderm both in vitro and in ovo. In: Nakamura H (ed) Electroporation and sonoporation in developmental biology. Springer, Tokyo, pp 73–83. https://doi.org/10.1007/978-4-431-09427-2_8

46. Schlueter J, Mikawa T (2018) Body cavity development is guided by morphogen transfer between germ layers. Cell Rep 24 (6):1456–1463. https://doi.org/10.1016/j.celrep.2018.07.015

47. Toy J, Bradford RL, Adler R (2000) Lipid-mediated gene transfection into chick embryo retinal cells in ovo and in vitro. J Neurosci Methods 104(1):1–8. https://doi.org/10.1016/S0165-0270(00)00311-3

48. deCastro M, Saijoh Y, Schoenwolf GC (2006) Optimized cationic lipid-based gene delivery reagents for use in developing vertebrate embryos. Dev Dyn 235(8):2210–2219. https://doi.org/10.1002/dvdy.20873

49. Cadinanos J, Bradley A (2007) Generation of an inducible and optimized piggyBac transposon system. Nucleic Acids Res 35(12):e87. https://doi.org/10.1093/nar/gkm446

50. Jordan BJ, Vogel S, Stark MR, Beckstead RB (2014) Expression of green fluorescent protein in the chicken using in vivo transfection of the piggyBac transposon. J Biotechnol 173:86–89. https://doi.org/10.1016/j.jbiotec.2014.01.016

51. Subramaniam A, Jones WK, Gulick J, Wert S, Neumann J, Robbins J (1991) Tissue-specific regulation of the alpha-myosin heavy chain gene promoter in transgenic mice. J Biol Chem 266(36):24613–24620

52. Agah R, Frenkel PA, French BA, Michael LH, Overbeek PA, Schneider MD (1997) Gene recombination in postmitotic cells. Targeted expression of Cre recombinase provokes cardiac-restricted, site-specific rearrangement in adult ventricular muscle in vivo. J Clin Invest 100(1):169–179. https://doi.org/10.1172/jci119509

53. Fisher SA, Watanabe M (1996) Expression of exogenous protein and analysis of morphogenesis in the developing chicken heart using an adenoviral vector. Cardiovasc Res 31:E86–E95

54. Watanabe M, Choudhry A, Berlan M, Singal A, Siwik E, Mohr S, Fisher SA (1998) Developmental remodeling and shortening of the cardiac outflow tract involves myocyte programmed cell death. Development 125(19):3809–3820

55. Dougherty JP, Temin HM (1986) High mutation rate of a spleen necrosis virus-based retrovirus vector. Mol Cell Biol 6(12):4387–4395. https://doi.org/10.1128/mcb.6.12.4387

56. Mikawa T, Fischman DA, Dougherty JP, Brown AM (1991) In vivo analysis of a new lacZ retrovirus vector suitable for cell lineage marking in avian and other species. Exp Cell Res 195(2):516–523. https://doi.org/10.1016/0014-4827(91)90404-i

57. Bell GW, Yatskievych TA, Antin PB (2004) GEISHA, a whole-mount in situ hybridization gene expression screen in chicken embryos. Dev Dyn 229(3):677–687. https://doi.org/10.1002/dvdy.10503

58. Darnell DK, Kaur S, Stanislaw S, Davey S, Konieczka JH, Yatskievych TA, Antin PB (2007) GEISHA: an in situ hybridization gene expression resource for the chicken embryo. Cytogenet Genome Res 117(1–4):30–35. https://doi.org/10.1159/000103162

59. Antin PB, Yatskievych TA, Davey S, Darnell DK (2014) GEISHA: an evolving gene expression resource for the chicken embryo. Nucleic Acids Res 42(Database issue):D933–D937. https://doi.org/10.1093/nar/gkt962

60. Ribatti D (2018) The chick embryo chorioallantoic membrane. In: Ribatti D (ed) In vivo models to study angiogenesis. Academic, Cambridge, MA, pp 1–23. https://doi.org/10.1016/B978-0-12-814020-8.00001-9

61. Dagg CP, Karnofsky DA, Roddy J (1956) Growth of transplantable human tumors in the chick embryo and hatched chick. Cancer Res 16(7):589–594

62. Auerbach R, Kubai L, Sidky Y (1976) Angiogenesis induction by tumors, embryonic tissues, and lymphocytes. Cancer Res 36(9 pt 2):3435–3440

63. Knighton D, Ausprunk D, Tapper D, Folkman J (1977) Avascular and vascular phases of tumour growth in the chick embryo. Br J Cancer 35(3):347–356. https://doi.org/10.1038/bjc.1977.49

64. Armstrong PB, Quigley JP, Sidebottom E (1982) Transepithelial invasion and intramesenchymal infiltration of the chick embryo chorioallantois by tumor cell lines. Cancer Res 42(5):1826–1837

65. Kunzi-Rapp K, Genze F, Kufer R, Reich E, Hautmann RE, Gschwend JE (2001) Chorioallantoic membrane assay: vascularized 3-dimensional cell culture system for human prostate cancer cells as an animal substitute model. J Urol 166(4):1502–1507. https://doi.org/10.1016/s0022-5347(05)65820-x

66. Durupt F, Koppers-Lalic D, Balme B, Budel L, Terrier O, Lina B, Thomas L, Hoeben RC, Rosa-Calatrava M (2012) The chicken chorioallantoic membrane tumor assay as model for qualitative testing of oncolytic adenoviruses. Cancer Gene Ther 19(1):58–68. https://doi.org/10.1038/cgt.2011.68

67. Ribatti D, Vacca A, Roncali L, Dammacco F (1996) The chick embryo chorioallantoic membrane as a model for in vivo research on angiogenesis. Int J Dev Biol 40(6):1189–1197

68. Hutson MR, Kirby ML (2003) Neural crest and cardiovascular development: a 20-year perspective. Birth Defects Res C Embryo Today 69(1):2–13. https://doi.org/10.1002/bdrc.10002

69. Bressan M, Davis P, Timmer J, Herzlinger D, Mikawa T (2009) Notochord-derived BMP antagonists inhibit endothelial cell generation and network formation. Dev Biol 326 (1):101–111. https://doi.org/10.1016/j.ydbio.2008.10.045

70. Wei Y, Mikawa T (2000) Formation of the avian primitive streak from spatially restricted blastoderm: evidence for polarized cell division in the elongating streak. Development 127 (1):87–96

71. Reese DE, Hall CE, Mikawa T (2004) Negative regulation of midline vascular development by the notochord. Dev Cell 6(5):699–708. https://doi.org/10.1016/s1534-5807(04) 00127-3

72. Finkelstein EB, Poole TJ (2003) Vascular endothelial growth factor: a regulator of vascular morphogenesis in the Japanese quail embryo. Anat Rec A Discov Mol Cell Evol Biol 272(1):403–414. https://doi.org/10. 1002/ar.a.10047

73. Nimmagadda S, Geetha Loganathan P, Huang R, Scaal M, Schmidt C, Christ B (2005) BMP4 and noggin control embryonic blood vessel formation by antagonistic regulation of VEGFR-2 (Quek1) expression. Dev Biol 280(1):100–110. https://doi.org/10. 1016/j.ydbio.2005.01.005

74. Bouvrée K, Larrivee B, Lv X, Yuan L, DeLafarge B, Freitas C, Mathivet T, Breant C, Tessier-Lavigne M, Bikfalvi A, Eichmann A, Pardanaud L (2008) Netrin-1 inhibits sprouting angiogenesis in developing avian embryos. Dev Biol 318(1):172–183. https://doi.org/ 10.1016/j.ydbio.2008.03.023

75. Mozdziak PE, Borwornpinyo S, McCoy DW, Petitte JN (2003) Development of transgenic chickens expressing bacterial beta-galactosidase. Dev Dyn 226(3):439–445. https://doi.org/10.1002/dvdy.10234

76. Chapman SC, Lawson A, Macarthur WC, Wiese RJ, Loechel RH, Burgos-Trinidad M, Wakefield JK, Ramabhadran R, Mauch TJ, Schoenwolf GC (2005) Ubiquitous GFP expression in transgenic chickens using a lentiviral vector. Development 132(5):935–940. https://doi.org/10.1242/dev.01652

77. Sato Y, Poynter G, Huss D, Filla MB, Czirok A, Rongish BJ, Little CD, Fraser SE, Lansford R (2010) Dynamic analysis of vascular morphogenesis using transgenic quail embryos. PLoS One 5(9):e12674. https://doi.org/10.1371/journal.pone.0012674

78. Sid H, Schusser B (2018) Applications of gene editing in chickens: a new era is on the horizon. Front Genet 9:456. https://doi.org/10. 3389/fgene.2018.00456

79. Sato Y, Lansford R (2013) Transgenesis and imaging in birds, and available transgenic reporter lines. Develop Growth Differ 55 (4):406–421. https://doi.org/10.1111/dgd. 12058

80. Komor AC, Badran AH, Liu DR (2017) CRISPR-based technologies for the manipulation of eukaryotic genomes. Cell 168 (1–2):20–36. https://doi.org/10.1016/j.cell. 2016.10.044

81. Woodcock ME, Idoko-Akoh A, McGrew MJ (2017) Gene editing in birds takes flight. Mamm Genome 28(7–8):315–323. https:// doi.org/10.1007/s00335-017-9701-z

82. Lee J, Ma J, Lee K (2019) Direct delivery of adenoviral CRISPR/Cas9 vector into the blastoderm for generation of targeted gene knockout in quail. Proc Natl Acad Sci U S A 116 (27):13288–13292. https://doi.org/10. 1073/pnas.1903230116

Chapter 10

Dynamic Imaging of Mouse Embryos and Cardiac Development in Static Culture

Andrew L. Lopez III and Irina V. Larina

Abstract

Dynamic imaging is a powerful approach to assess the function of a developing organ system. The heart is a dynamic organ that undergoes quick morphological and mechanical changes through early embryonic development. Defining the embyonic mouse heart's normal function is important for our own understanding of human heart development and will inform us on treatments and prevention of congenital heart defects (CHD). Traditional methods such as ultrasound or fluorescence-based microscopy are suitable for live dynamic imaging, are excellent to visualize structure and connect gene expression to phenotypes, but can be of low quality in resolving fine features and lack imaging depth and scale to fully appreciate organ morphogenesis. Additionally, previous methods can be limited in accommodating a live imaging apparatus capable of sustaining whole embryo development for extended periods time. Optical coherence tomography (OCT) is unique in this circumstance because acquisition of three-dimensional images without contrast reagents, at single cell resolution make it a suitable modality to visualize fine structures in the developing embryo. OCT setups are highly customizable for live imaging because of the tethered imaging arm, due to its setup as a fiber-based interferometer. OCT allows for 4D (3D + time) functional imaging of living mouse embryos and can provide functional and mechanical information to ascertain how the heart's pump function changes through development. In this chapter, we will focus on how we use OCT to visualize live heart dynamics at different stages of development and provide mechanical information to reveal functional properties of the developing heart.

Key words Optical coherence tomography, Cardiovascular development, Embryo culture, Cardio-dynamic analysis, Mouse, Live imaging, Heart morphogenesis

1 Introduction

The vertebrate embryonic heart is the first functional organ in development, and is vital to support the growing nutritive and respiratory demands of the developing embryo. To propel growth and development, the primitive heart executes morphological and mechanical changes to achieve proper configuration, increase cardiac output, and maintain respiratory homeostasis. The heart's failure to execute these changes implicates developmental challenges ranging from embryonic death to congenital heart defects;

Domenico Ribatti (ed.), *Vascular Morphogenesis: Methods and Protocols*, Methods in Molecular Biology, vol. 2206, https://doi.org/10.1007/978-1-0716-0916-3_10, © Springer Science+Business Media, LLC, part of Springer Nature 2021

such defects can mildly or severely impact post utero development and the ability to thrive. In the USA, eight in 1000 live births incur a congenital heart defect, according to the American Heart Association [1]. Therefore, it is important to investigate all aspects of cardiovascular development, from gene regulation to mechanical properties of the heart, to develop the next generation of prophylactic treatments and therapies to improve outcomes in patients with congenital heart defects.

Our current understanding of embryonic cardiovascular development is bolstered by strides in molecular biology, development of tomographic and fluorescence-based imaging, and well-established histological analysis. These approaches have generated a foundational collection of work on gene programs and cell-biology that regulate normal development, and to this day provide insight on the pathology of heart disease. While powerful and informative, these approaches lack functional, in vivo analysis that can describe how biomechanics influence cardiogenesis. In fact, the heart's pump function and generation of hemodynamic load are more regularly being assessed as moderators of gene activation and development, but progress is hindered by the challenge of experimentation with mechanical functions in a live embryonic heart [2–7]. To this end, our group is focused on developing live embryo culture methods coupled with advanced imaging modalities, such as optical coherence tomography (OCT), to unveil information about the mechanical environment that regulates development in the embryonic heart [8–10].

In this chapter, we present the mouse embryo as a model for mammalian cardiovascular development. We will describe live embryo culture methods for OCT imaging to analyze live embryos, assess cardiodynamics, and quantify hemodynamics in the developing heart.

2 Materials

2.1 Embryo-Dissection Medium

1. DMEM/F12 with 2.5 mM L-Glutamine and 15 mM HEPES buffer (Thermo Fisher Scientific Inc., Waltham, MA).

2. Fetal Bovine Serum (FBS) (Gibco, Carlsbad, CA), stored in 5 mL aliquots at −30 °C.

3. Penicillin/streptomycin, 100× (Gibco, Carlsbad, CA), stored as 5 mL aliquots at −30 °C.

2.2 Rat Serum Extraction

1. Adult male Sprague Dawley rats (Charles Rivers Laboratories, Wilmington, MA).

2. Ether (J.T. Baker Phillipsburg, NJ).

3. Vacutainer blood collection tubes (BD Biosciences, Franklin Lakes, NJ).

4. Vacutainer blood collection sets (BD Biosciences, Franklin Lakes, NJ).

5. 0.45 μm syringe filter (Nalgene).

2.3 Embryo Dissection

1. Two #5 micro dissecting forceps (Roboz Surgical Instrument Co., Inc., Gaithersburg, MD).

2. Micro dissecting scissors (Roboz Surgical Instrument Co., Inc., Gaithersburg, MD).

3. Microdissecting tweezers (Roboz Surgical Instrument Co., Inc., Gaithersburg, MD).

4. 35 mm petri dishes.

2.4 Embryo-Culture Medium

1. DMEM/F12 with 2.5 mM L-glutamine and 15 mM HEPES buffer (HyClone, Logan, UT).

2. Rat serum (stored as 1 mL aliquots at −80 °C).

3. Penicillin/streptomycin, 100× (Gibco, Carlsbad, CA), stored as 5 mL aliquots at −30 °C.

3 Methods

3.1 Rat Serum Extraction (See Note 1)

1. Anesthetize rat with ether (*see* **Note 2**).

2. Expose dorsal aorta by abdominal incision with scissors.

3. Collect blood into a vacutainer blood collection tube using the vacutainer blood collection set.

4. Once blood is collected, invert the tube a few times to mix and keep it on ice while extracting blood from other rats.

5. Centrifuge tubes with collected blood at $1300 \times g$ for 20 min. The supernatant should look golden. If the supernatant looks pink, which indicates erythrocyte lysis, we recommend discarding those tubes. Pool supernatant (serum) into 15 mL centrifuge tubes.

6. Centrifuge serum at $1300 \times g$ for 10 min to pellet down the remaining erythrocytes. Pool serum into 50 mL centrifuge tubes.

7. Heat-inactivate serum at 56 °C for 30 min in a water bath. Keep tube lid unscrewed for ether evaporation.

8. Keep serum overnight at 4 °C with the lid unscrewed for further ether evaporation.

9. Filter serum through a 0.45 μm syringe filter.

10. Aliquot serum by 1 mL and keep at −80 °C.

3.2 Embryo Dissections for Live Imaging

1. Set up a mating cage and check for vaginal plugs every morning. If a plug is found, the pregnancy is recorded as E0.5.

2. At the desired embryonic stage, prepare dissection medium with 89% DMEM/F-12, 10%FBS, and 1% 100× penicillin/streptomycin. Approximately 100 mL of medium will be used per litter.

3. Preheat dissection medium to 37 °C.

4. Preheat the dissection station and maintain at 37 °C. We use a custom-built heating station (Fig. 1a) composed of a plexiglass chamber surrounding a dissection microscope, a commercial space heater, and a temperature controller.

5. Sacrifice the pregnant mouse in accordance with established animal-use guidelines.

6. Dissect the uterus with embryos and transfer over to warm (37 °C) dissection medium in a petri dish. Immediately move the dish into the temperature-controlled dissection station. Dissection procedures are detailed in Lopez et al. [11].

7. With microdissecting forceps, under microscope guidance, sequentially separate individual embryos from the muscular uterine tissue.

8. Delicately peel away the deciduum covering the embryo. Leave large piece of the deciduum connected to the ectoplacental cone. This remaining tissue can be shaped with dissection forceps to position the embryo with the heart toward the imaging beam.

9. Delicately peel away Reichert's membrane exposing the yolk sac, while leaving the yolk sac intact (*see* **Note 3**).

10. Transfer the embryo to a culture dish with fresh warm media using a transfer pipette.

11. For short-term experiments limited to a few hours (e.g., for live analysis of cardiodynamics), embryos can be imaged in dissection media. However, for longer imaging sessions, when embryos are expected to progress through development, the culture medium should be supplemented with rat serum: two parts DMEM/F12, one-part rat serum, and 1% 100× penicillin/streptomycin.

12. Place culture dish the embryo for recovery in a 5% CO_2 and 37 °C incubator for 30 min to 1 h.

3.3 Imaging Setup

The application of OCT imaging is advantageous because it relies on the light scattering property of tissue. In contrast, popular imaging modalities use immunochemistry or contrast reagents that greatly increase experimental turnover time and costs. Different OCT systems exist for embryo imaging and are designed to

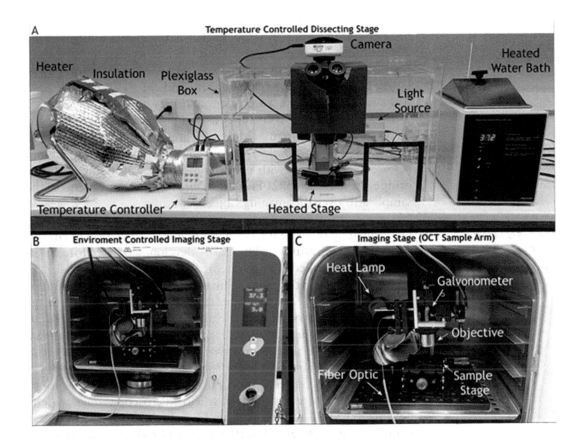

Fig. 1 Dissection and imaging stage. (a) Embryos are dissected in a heated lab-built dissection stage consisting of a plexiglass box, an arduino controlled heated stage, and a commercial heater connected to the plexiglass box and modulated by a temperature controller. (b) Dissected embryos are imaged in a commercial incubator maintaining 37 °C and 5% CO_2. (c) OCT sample arm resides within the incubator to image live embryos. Heating lamp is used to maintain heat while incubator is open

meet the purpose of the experiment. For our purpose, we built a spectral-domain OCT system with the ability to generate broad bandwidth light for high resolution imaging; produce sufficient power for excellent contrast and imaging depth, while not damaging a live specimen; and operate at an A-line acquisition rate up to 250 kHz. To live image, our OCT imaging arm is designed to fit a commercial, environment-controlled incubator (Fig. 1b, c) (*see* **Notes 4** and **5**). Our OCT system is based on the MICRA 5 Titanium-Sapphire laser (Coherent Inc.) with a central wavelength of 808 nm and a bandwidth of ~110 nm [8]. At a 1D imaging rate of ~68 kHz, we obtain a spatial resolution of ~5 μm in the axial direction (in tissue) and ~4 μm in the transverse direction. The system sensitivity was measured at ~97 dB with a sensitivity drop of ~4 dB over ~1 mm in depth.

A 1D depth profile (A-scan) is obtained by performing a fast Fourier transform on fringe data that is transformed into equally sampled data in linear k-space using the laser spectra information. Fringe data is obtained by fiber-based Michelson interferometry of light reflected from the reference arm and the back reflected light from the sample arm. Interference fringes are spatially resolved using a lab-built spectrometer consisting of a diffraction grating, a focusing lens system, and a line-field CMOS camera (spL4096–140 km, Basler). 2D images (B-scan) and 3D volumes (C-scan) are obtained with an x-axis and y-axis galvanometer (Cambridge Technology 6220H with 8 mm mirror) mounted on the sample arm over the imaging objective.

3.4 Visualization of Embryonic Structures Using OCT

OCT imaging of live static culture provides visualization of embryonic structural features through early development (*see* **Note 6**). In a climate-controlled environment, embryos cultured at E7.5–E10.5 stages reach the same hallmarks of development as embryos in utero [8, 9, 12]. This becomes especially important when imaging mutant embryos, which are usually less robust and need an adequate environment to survive. As an example, Figure 2 shows a structural visual-analysis of embryos at E8.5 and E9.5 with OCT in the Wdr19 mouse model. Wdr19 mutants are embryonic lethal at E10.5 and exhibit defects in neural tube closure and cardiac looping [8]. The figure shows bright-field microscopic images (left) next to 3D OCT reconstructions (right). Wild Type embryos as controls (Fig. 2a, b) and Wdr19 mutant embryos (Fig. 2c, d) at two developmental stages.

Size difference in whole embryos is apparent, control embryos at E9.5 demonstrate normal neural tube development, and Wdr19 embryos exhibit aberrant neural tube closure resulting in a severe craniofacial defect.

Cellular level resolution allows for visualization and analysis of structural defects in embryonic mouse mutants. But for later stage embryos, the imaging depth of the OCT becomes insufficient to visualize internal structures with any detail. In this case, embryos can be dissected out of the yolk sac for imaging. While this approach is not compatible with live imaging, it reveals structural features at a cellular level without application of contrast agents in freshly dissected embryos, which might be beneficial in a variety of applications.

3.5 Cardiodynamic Analysis

A high volume acquisition rate is critical for cardiodynamic analysis. Nonetheless, traditional OCT systems cannot achieve the ~100 frames per second frame rate required for direct volumetric acquisition and heart wall tracking. To resolve this limitation, a number of algorithms have been proposed to synchronize time sequences acquired over multiple heartbeats at neighboring positions through the entire heart volume [13–20]. These methods rely on the

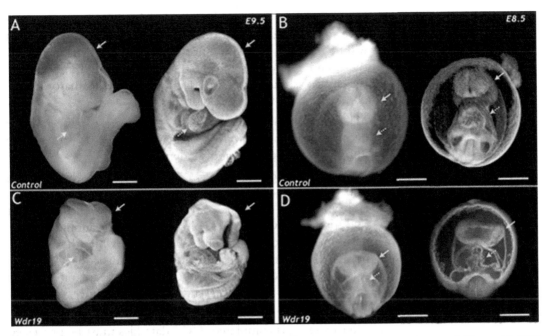

Fig. 2 Structural analysis of early stage mouse embryos. Brightfield microscopic images (left) and 3D OCT images (right) of control embryos (**a, b**) and Wdr19 mutant (**c, d**) embryos at E9.5 (**a + c**) and E8.5 (**b + d**). Solid arrows point at the neural tube in the head region and dashed arrows point at the heart. Scale bars correspond to 500 μm. (Adapted from [8])

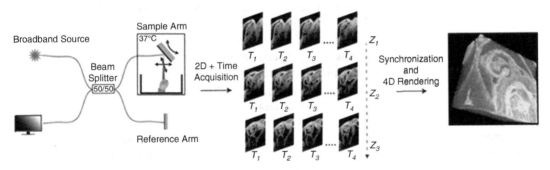

Fig. 3 Volumetric cardiodynamic imaging approach. An illustration depicting our OCT set and 4D acquisition algorithm. (Adapted from [8])

periodic nature of the cardiac cycle. Given below is an algorithm based on nongated synchronization (Fig. 3), which we successfully used for reconstruction of 4D beating hearts [8, 12]. This approach is best for embryos from E8.5 to E9.5.

1. Position the embryo in the culture dish so that the heart region of the embryo faces up toward the imaging beam. Leaving a piece of deciduum attached to the ectoplacental cone will allow one to manipulate the tissue into a base and keep the embryo steady.

2. Once the embryo is positioned, spatial acquisition parameters have to be assigned to assure that the entire volume of heart will be imaged. For E8.5 embryos, our scanning amplitude in the X and Y direction is ~750 ± 150 μm. From this point on, the embryo should remain undisturbed in the incubator for the duration of the imaging session.

3. Make note of the assigned spatial parameters, as they will be necessary for computing the spatial geometry of your sample during post processing and rendering.

4. Assign scanning parameters. As a suggestion, we assign 600 A-lines and 20,000 B-scans per volume for an E8.5 embryo. The B-scans are continuously acquired at 100 frames/s over a single volume (the Y-galvanometer scanner is slowly moving through the volume while the X-galvanometer scanner is quickly moving for B-scan acquisition). As a result, during each heartbeat cycle (the heart rate of the embryo is ~2 Hz at this stage) the Y scanner moves by ~2 μm, which is below the lateral resolution of the system.

5. After imaging is completed, OCT data files must be processed into an image sequence of image files corresponding to B-scans for processing and visualization.

6. Define the number of frames per heartbeat cycle. Split the total data set into individual cycles and assign the position for each heartbeat cycle according to the total scan size in the Y direction and the total acquired number of heartbeats. We use Imaris v8.0.2 software (Bitplane, Switzerland) for 3D rendering and processing. The heartbeats are synchronized to the same phase of the heartbeat cycle using previously described algorithms [21], revealing the 4D (3D plus time) cardiodynamic data.

3.6 Imaging the Vasculature Using Speckle Variance

Vascular morphology of the yolk sac and the embryo proper can be visualized using speckle variance (SV) OCT approach. This method relies on the analysis of temporal variations in pixel intensity in structural OCT images. Interaction of laser radiation with tissue scatterers creates a speckle pattern. The SV OCT method segments out varying pixel intensities due to moving scatterers at specific positions. Since most moving scatterers in embryos are blood cells, this method can be used to reveal vascular structure [22, 23]. Speckle variance images can be generated using the following formula, which determines the variance between N images in time, t:

$$SV_{i,j} = \frac{1}{N} \sum_{t=0}^{N-1} \left(I_{i,j,t} - I_{i,j,\text{mean}} \right)^2$$

where, $SV_{i,j}$ is the variance at a specific pixel (i,j), $I_{i,j,t}$ is the structural image intensity at that pixel at time t, and $I_{i,j,\mathrm{mean}}$ is the mean intensity of pixel (i,j) for all N values [22]. Speckle variance approach provides vascular structure in the developing embryo and extraembryonic structures; however, movements associated with the beating embryonic heart can degrade the SV signal, and therefore, limits it to nonmoving vascular beds.

3.7 Live Imaging of Hemodynamics Using Doppler OCT

Quantitative blood flow analysis is an important component to functionally assess the heart. Doppler OCT relies on the measurement of OCT phase shifts between consecutive A-scans at each pixel and can be used to quantify hemodynamics at the same spatial and temporal resolution as structural OCT imaging [24]. The direction of flow relative to the laser beam needs to be defined for quantitative blood flow analysis. The flow velocity at each pixel can be reconstructed using the formula [25]:

$$\nu = \frac{\Delta\varphi}{2n\frac{2\pi}{\lambda}\tau\cos\beta}$$

where φ is a Doppler phase shift between adjacent A-scans, n is the refractive index, λ is the central wavelength used for imaging, τ is the time between consecutive A-scans, and β is the angle between the flow direction and the laser beam. The angle β is calculated from structural 2D and 3D data sets acquired from imaged embryos. To image embryonic mice we usually assume a refractive index of $n = 1.4$.

The resolution of Doppler OCT allows for quantitative hemodynamic analysis from bulk blood cell movement [26] as well as individual circulating erythrocytes at early circulation stages, when the majority of blood cells are still restricted from flowing in the blood islands [27]. Spatially and temporally resolved blood flow profiles can be reconstructed from these measurements. Hemodynamic measurements acquired by Doppler OCT in cultured mouse embryos are comparable to measurements made using fast-scanning confocal microscopy in superficial vessels. However, flow measurements deep within the embryo are not possible with confocal microscopy due to the imaging depth limitation of this method, and unique to the Doppler OCT technique [26, 28].

Volumetric blood flow analysis in the beating heart with Doppler OCT can be performed using the synchronization approach described above for structural imaging [12]. For that, structural and Doppler OCT images are exported and treated as two channels of the same data set. The synchronization is performed based on the structural data set, while the Doppler OCT frames are passively rearranged and registered based on the corresponding structural data arrangements. Figure 4 shows an example of volumetric Doppler OCT analysis of blood flow in the embryonic heart. The

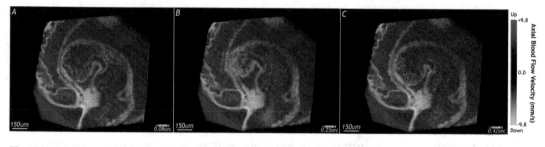

Fig. 4 (a–c) Volumetric hemodynamic imaging in the embryonic heart. Snapshots from 4D hemodynamic imaging of a late E8.5 wild type embryonic mouse heart at different phases of the cardiac cycle. The structural data is shown in gray scale while the Doppler OCT velocity measurement is overlaid in color

blood flow labeled with blue (toward the detector) and magenta (away from the detector) is clearly distinguishable in the heart and extraembryonic vessels at different phases of the heartbeat cycle.

Analysis of absolute blood flow velocity in the heart might be challenging due to difficulty of evaluating the flow direction. At early heart tube stages, when the flow is laminar and the direction of flow is obvious based on the 3D structure of the heart tube, the absolute flow velocity profiles in the heart can be reconstructed and are comparable with ultrasonic measurements acquired in utero [12]. This provides an opportunity for detailed quantitative volumetric biomechanical analysis in the embryonic heart by correlating heart wall dynamics with local flow measurement [12]. An example of correlative analysis of the cardiodynamics and hemodynamics is shown in Fig. 5. This can reveal the details of the cardiac cycle previously not known and opens a door for studies on biomechanical regulation of cardiac function.

3.8 Conclusion

Here we describe methods for dynamic imaging of the developing embryonic mouse heart. Using OCT and postprocessing methods, one can extract detailed structural and functional information. This is highly relevant when trying to understand how dynamic and mechanical changes influence development and resemble mutant phenotypes, in order to connect gene activity with mechanical signaling. Combined with molecular genetic approaches these methods can help us generate a more complete view of all the parameters that regulate development.

4   Notes

1. Rat serum is available commercially; however, we prepare our own serum via the protocol provided here.

2. We use ether for rat serum extraction because it evaporates from the serum unlike other anesthetics.

3. We find it helpful to change out the media during the dissection to assure pH and temperature are consistent.

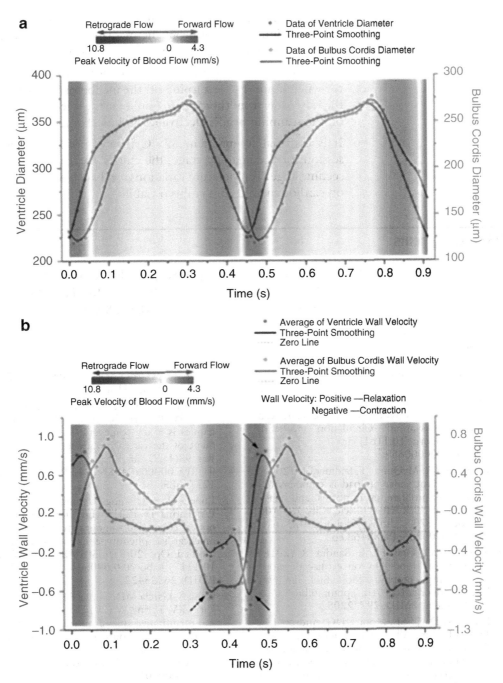

Fig. 5 Analysis of hemodynamics in relation to heart wall dynamics. (**a**) Diameters and (**b**) average heart wall velocities of the ventricle and the bulbus cordis are plotted with blood flow dynamics in the bulboventricular region of a Wild Type E9.0 embryo. The heartbeat is duplicated for a second cycle for better visualization. (Adapted from [12])

4. We recommend tilting the scanning head slightly (~5°) to minimize light reflection from the media surface and the dish bottom, which might significantly improve the imaging quality.

5. We recommend removing the scanning head from the incubator when not in use or turning off the incubator and removing the water tray from the stage to minimize the effect of moisture on the scanning head's electronics and mechanics.

6. It is essential to maintain 37 °C through out the imaging session for the imaging stage, the culture media, and the dissecting stage. Temperature variations can have a dramatic effect on cardiodynamics and embryo viability.

Acknowledgments

This work is supported by grants from the National Institute of Health R01HL120140, R01EB027099, and R01HD096335 and the American Heart Association 19PRE34380240.

References

1. Benjamin EJ, Muntner P, Alonso A, Bittencourt MS, Callaway CW, Carson AP et al (2019) Heart disease and stroke statistics—2019 update: a report from the American Heart Association. Circulation 139:e56. https://doi.org/10.1161/cir.0000000000000659

2. deAlmeida A, McQuinn T, Sedmera D (2007) Increased ventricular preload is compensated by myocyte proliferation in normal and hypoplastic fetal chick left ventricle. Circ Res 100 (9):1363–1370. https://doi.org/10.1161/01.RES.0000266606.88463.cb

3. Katherine C, Graham R, Sandra R (2018) Influence of blood flow on cardiac development. Prog Biophys Mol Biol 137:95. https://doi.org/10.1016/j.pbiomolbio.2018.05.005. PMID: 29772208

4. Shang W, David SL, Monica DG, Andrew LL, Kirill VL, Irina VL (2016) Four-dimensional live imaging of hemodynamics in mammalian embryonic heart with Doppler optical coherence tomography. J Biophotonics 9 (8):837–847. https://doi.org/10.1002/jbio.201500314. PMID: 26996292

5. Michael CB, Jonathan DL, Takashi M (2014) Hemodynamic forces regulate developmental patterning of atrial conduction. PLoS One 9 (12):e115207. https://doi.org/10.1371/journal.pone.0115207. PMID: 25503944

6. Madeline M, Sandra R (2014) Congenital heart malformations induced by hemodynamic altering surgical interventions. Front Physiol 5:287. https://doi.org/10.3389/fphys.2014.00287. PMID: 25136319

7. Maria R, Carlin R, Angela D, Chiffvon PS, Andy W, Robert GG et al (2003) Hemodynamics is a key epigenetic factor in development of the cardiac conduction system. Circ Res 93(1):77–85. https://doi.org/10.1161/01.RES.0000079488.91342.B7. PMID: 12775585

8. Lopez AL 3rd, Wang S, Larin KV, Overbeek PA, Larina IV (2015) Live four-dimensional optical coherence tomography reveals embryonic cardiac phenotype in mouse mutant. J Biomed Opt 20(9):90501. https://doi.org/10.1117/1.jbo.20.9.090501. PubMed PMID: 26385422

9. Wang S, Garcia MD, Lopez AL, Overbeek PA, Larin KV, Larina IV (2017) Dynamic imaging and quantitative analysis of cranial neural tube closure in the mouse embryo using optical coherence tomography. Biomed Opt Express 8(1):407–419. https://doi.org/10.1364/BOE.8.000407

10. Lopez AL 3rd, Larina IV (2019) Second harmonic generation microscopy of early embryonic mouse hearts. Biomed Opt Express 10 (6):2898–2908. https://doi.org/10.1364/BOE.10.002898. PubMed PMID: 31259060

11. Lopez AL 3rd, Garcia MD, Dickinson ME, Larina IV (2015) Live confocal microscopy of the developing mouse embryonic yolk sac vasculature. Methods Mol Biol 1214:163–172.

https://doi.org/10.1007/978-1-4939-1462-3_9. PubMed PMID: 25468603

12. Wang S, Lakomy DS, Garcia MD, Lopez AL, Larin KV, Larina IV (2016) Four-dimensional live imaging of hemodynamics in mammalian embryonic heart with Doppler optical coherence tomography. J Biophotonics 9 (8):837–847. https://doi.org/10.1002/jbio. 201500314

13. Liu A, Wang R, Thornburg KL, Rugonyi S (2009) Efficient postacquisition synchronization of 4-D nongated cardiac images obtained from optical coherence tomography: application to 4-D reconstruction of the chick embryonic heart. J Biomed Opt 14(4):044020. https://doi.org/10.1117/1.3184462. PubMed PMID: 19725731

14. Jenkins MW, Rothenberg F, Roy D, Nikolski VP, Hu Z, Watanabe M et al (2006) 4D embryonic cardiography using gated optical coherence tomography. Opt Express 14 (2):736–748. PubMed PMID: 19503392

15. Jenkins MW, Chughtai OQ, Basavanhally AN, Watanabe M, Rollins AM (2007) In vivo gated 4D imaging of the embryonic heart using optical coherence tomography. J Biomed Opt 12 (3):030505. https://doi.org/10.1117/1. 2747208. PubMed PMID: 17614708

16. Mariampillai A, Standish BA, Munce NR, Randall C, Liu G, Jiang JY et al (2007) Doppler optical cardiogram gated 2D color flow imaging at 1000 fps and 4D in vivo visualization of embryonic heart at 45 fps on a swept source OCT system. Opt Express 15(4):1627–1638. PubMed PMID: 19532397

17. Larin KV, Larina IV, Liebling M, Dickinson ME (2009) Live imaging of early developmental processes in mammalian embryos with optical coherence tomography. J Innov Opt Health Sci 2(3):253–259. https://doi.org/10.1142/s1793545809000619. PubMed PMID: 20582330; PubMed Central PMCID: PMCPMC2891056

18. Gargesha M, Jenkins MW, Wilson DL, Rollins AM (2009) High temporal resolution OCT using image-based retrospective gating. Opt Express 17(13):10786–10799. PubMed PMID: 19550478; PubMed Central PMCID: PMCPMC2748662

19. Daniels K, Daugherty J, Jones J, Mosher W (2015) Current contraceptive use and variation by selected characteristics among women aged 15–44: United States, 2011–2013. Natl Health Stat Rep 86:1–14

20. Division of Reproductive Health, National Center for Chronic Disease Prevention and Health Promotion, Centers for Disease Control and Prevention (CDC) (2015) U.S. Selected Practice Recommendations for Contraceptive Use, 2013: adapted from the World Health Organization selected practice recommendations for contraceptive use, 2nd edition. MMWR Recomm Rep 62 (RR-05):1–60

21. Liebling M, Forouhar AS, Gharib M, Fraser SE, Dickinson ME (2005) Four-dimensional cardiac imaging in living embryos via postacquisition synchronization of nongated slice sequences. J Biomed Opt 10(5):054001. https://doi.org/10.1117/1.2061567. PubMed PMID: 16292961

22. Sudheendran N, Syed SH, Dickinson ME, Larina IV, Larin KV (2011) Speckle variance OCT imaging of the vasculature in live mammalian embryos. Laser Phys Lett 8(3):247–252

23. Mariampillai A, Leung MK, Jarvi M, Standish BA, Lee K, Wilson BC et al (2010) Optimized speckle variance OCT imaging of microvasculature. Opt Lett 35(8):1257–1259. https://doi.org/10.1364/ol.35.001257. PubMed PMID: 20410985

24. Chen Z, Milner TE, Srinivas S, Wang X, Malekafzali A, van Gemert MJ et al (1997) Noninvasive imaging of in vivo blood flow velocity using optical Doppler tomography. Opt Lett 22(14):1119–1121. PubMed PMID: 18185770

25. Vakoc B, Yun S, de Boer J, Tearney G, Bouma B (2005) Phase-resolved optical frequency domain imaging. Opt Express 13 (14):5483–5493. PubMed PMID: 19498543; PubMed Central PMCID: PMCPMC2705336

26. Larina IV, Sudheendran N, Ghosn M, Jiang J, Cable A, Larin KV et al (2008) Live imaging of blood flow in mammalian embryos using Doppler swept-source optical coherence tomography. J Biomed Opt 13(6):060506. https://doi.org/10.1117/1.3046716. PubMed PMID: 19123647

27. Larina IV, Ivers S, Syed S, Dickinson ME, Larin KV (2009) Hemodynamic measurements from individual blood cells in early mammalian embryos with Doppler swept source OCT. Opt Lett 34(7):986–988. PubMed PMID: 19340193; PubMed Central PMCID: PMCPMC2874199

28. Jones EA, Baron MH, Fraser SE, Dickinson ME (2004) Measuring hemodynamic changes during mammalian development. Am J Physiol Heart Circ Physiol 287(4):H1561–H1569. https://doi.org/10.1152/ajpheart.00081. 2004. PubMed PMID: 15155254

## Chapter 11

High-Resolution Confocal Imaging of Pericytes in Human Fetal Brain Microvessels

Mariella Errede, Francesco Girolamo, and Daniela Virgintino

Abstract

Pericytes are integral part of neurovascular unit and play a role in the maintenance of blood-brain barrier integrity, angiogenesis, and cerebral blood flow regulation. Despite their important functional roles, a univocal phenotypic identification is still emerging also for the lack of a "pan-pericyte" marker. In the present study, we describe in detail the method for performing fluorescence immunohistochemistry on thick free-floating sections from human fetal brain in high resolution laser confocal microscopy. This method enables to obtain three-dimensional images of pericytes and provides insights about their distribution and localization in the microvessels of human developing brain.

Key words Pericyte, Vibratome sectioning, High-resolution confocal microscopy

1 Introduction

Pericytes are spatially isolated cells associated with the blood vessels closest to endothelial cells (ECs) and embedded within the vascular basal lamina (VBL) [1]. Pericytes are characterized by a round nucleus that protrudes from the vessel profile [2] and processes extending from the soma along and around precapillary arterioles, capillaries, and postcapillary venules [3]. In the central nervous system (CNS), pericytes take integral part in the neurovascular unit (NVU), an integrative biological system composed of endothelial cells, pericytes, vascular smooth muscle cells, astrocytes, microglia cells, and neurons [4] The interaction among all these cell types, by direct contact and through paracrine signaling pathways, contributes to the regulation of central nervous system homeostasis, and in particular the maintenance of the integrity of the blood–brain barrier (BBB) [1]. During brain development, pericytes play a leading role in angiogenesis in which they precede and guide the ECs growth [5–7], regulate the number of endothelial tight junctions (TJs) and polarize perivascular astrocyte endfeet, controlling de facto BBB and NVU functions [8]. In adult brain,

Domenico Ribatti (ed.), *Vascular Morphogenesis: Methods and Protocols*, Methods in Molecular Biology, vol. 2206, https://doi.org/10.1007/978-1-0716-0916-3_11, © Springer Science+Business Media, LLC, part of Springer Nature 2021

pericytes, although being quiescent, seem to undergo to a structural remodelling, contributing to endothelial stabilization [9]. During neurological diseases, pericytes undergo phenotypic and functional modifications and their dysfunctions or loss are to be considered involved in the pathogenesis of neurodegenerative diseases, stroke and brain tumors [6, 10, 11]. Since from their discovery, there has been debate about pericyte heterogeneous morphology depending on their localization in the CNS vascularization territories [12]. The majority of brain vasculature is composed of capillaries/microvessels [13], site of the BBB [14]; here, pericytes are organized as a cellular chain on the vessel length where they lay down their rounded cell bodies from which primary processes rise [15]. Hence, the morphology of pericyte would appear quite heterogeneous. At capillary level, the pericyte processes appear large, covering a wide surface of vessel [16]. Alternatively, these processes form fingerlike projections, thin and long, from which secondary perpendicular processes rise, partially encircling the vessels with tips closely adherent to ECs [3, 17]. The anatomical classification of brain pericytes could be reinforced by immunohistochemical staining thanks to the availability of many markers [18], even though a "pan-pericyte" marker has not yet been discovered; thus, the identification of pericytes by specific markers is influenced by species, tissue of origin and function, vessel type, developmental stage, and pathological conditions [13, 19, 20]. The most widely used markers to identify the pericytes are platelet-derived growth factor receptor beta (PDGFRβ), involved in pericytes recruitment and proliferation [20]; alpha smooth muscle actin (αSMA) [21], desmin and vimentin [22] related to pericytes contractile activity; CD146, known as melanoma cell adhesion molecule, implicated in cell communication during the BBB development [23]; CD13/aminopeptidase N, as surface marker [24]; chondroitin sulfate proteoglycan4/neural glial antigen 2 (NG2) expressed during vasculogenesis and angiogenesis [25]. NG2 is a transmembrane proteoglycan with a short cytoplasmic tail and large extracellular domain [26] characterized by numerous, variable sites of glycosylation giving rise to multiple array of glycoforms/isoforms able to exert variable-specific functions [27]. The availability of a large panel of antibodies against each specific glycoforms of NG2 allows to identify subsets of activated pericyte associated with growing vessels in fetal brain in high-resolution confocal microscopy [28]. The existence of pericyte subpopulations suggests their different functional/physiological roles in angiogenesis, BBB formation and maintenance, cerebral blood flow regulation, and leukocyte trafficking [8, 29]. In conclusion, pericytes represent a multifunctional cell population for whose identification different methods are required to better define their involvement in physiological and pathological conditions of the CNS. Here, we provide a detailed method for a three-dimensional outlook of pericytes distribution in the complex architecture of NVU based on high resolution confocal microscopy immunofluorescence on free-floating sections of human developing brain using a NG2 isoform. The protocol recommends a lighter fixation that must include a very small amount of glutaraldehyde, which consents to detect the fine pericyte morphology within the microvessels. Vibrating microtome (vibratome) sectioning is a straightforward technique particularly useful for revealing histological and 3D detail in tissues because does not require any harsh components maintaining antigen integrity suitable for immunofluorescent staining. In addition, immunohistochemistry applied on thick free-floating sections allows a double-sided antibody penetration, making it a more practical method for looking at the staining distribution through an entire brain region, or imaging structures through the depth of a section. The operating mode of the laser confocal microscope Leica makes it possible to fine-tune laser, fluorescence, and optical parameters and creates custom-made "Instrument Parameter Settings" for the highest optical resolution and the lowest diffraction.

2   Materials

Autopsy specimens of brain were obtained from fetus derived from spontaneous abortions at 22 weeks of gestation with no history of neurological pathologies. The sampling and handling of human specimens were conformed to the ethical rules of the Department of Pathology, Medical School, University of Bari, Italy, and approval was gained from the local Ethics Committee of the National Health System in compliance with the principles stated in the Declaration of Helsinki. The fetal age was evaluated on the basis of the crown–rump length and/or pregnancy records (counting from the last menstrual period). The dorsolateral walls of the telencephalic vesicles (future cortex of cerebral hemispheres) were dissected in samples of ≤0.5 cm in thickness.

3   Methods

3.1   Fixation and Sectioning

The high quality of fixation of brain samples is achieved by shortening the handling timing as described below.

1. Fixation by immersion in 2% paraformaldehyde plus 0.2% glutaraldehyde in phosphate-buffered saline (PBS; Sigma-Aldrich, Darmstadt, Germany) freshly prepared solution, for 3 h at 4 °C.

2. Storage of samples into 0.02% paraformaldehyde (0.02% PFA) in PBS buffer at 4 °C until sectioning.

3. Cutting of the sample in a regular-shape dowel by forceps under a stereomicroscope.

4. Mounting on the vibratome specimen holder (Leica Microsystem; Milton Keynes, UK) using cyanoacrylate glue allowing sufficient time for glue solidification.

5. Transfer of the specimen holder into vibratome buffer container filled with cold PBS until to completely submerge the sample.

6. Lock a new single-edge razor blade in the blade holder with the sharpened surface facing upward. Set the sectioning window around the specimen using the manufacturer's instructions.

7. Cut of 20-µm sections; the microtome is operated using the following adjustable settings: knife angle 5°; sectioning speed 0.05 mm/s; oscillation frequency, 100 Hz; oscillation amplitude, 0.6 mm.

8. Transfer each section using a thin rod of glass or a brush into a well containing cold 0.02% PFA in PBS buffer and store at 4 °C until use (it represents a remarkably resilient method of specimen and antigen preservation).

Once you have completed the specimen cutting, mount the slices onto slides for histological evaluation.

3.2 Immunofluorescence Staining

Transfer of the sections into a multiwall of 24-well plate filled with 300 µL PBS; each step was carried out transferring the sections to a clean well for each new solution. The sections undergo the following steps.

1. Rinse three times with 300 µL of PBS for 5 min at room temperature (RT) on a shaking platform.

2. Permeabilization with 0.5% Triton X-100 (Sigma-Aldrich) in PBS for 30 min at RT on a shaking platform.

3. Rinse three times with 300 µL of PBS for 5 min at RT on a shaking platform.

4. Blocking of nonspecific binding sites by incubation with blocking buffer-PBS (BB; 1% bovine serum albumin, 2% FCS in PBS) for 20 min at RT on a shaking platform.

5. Incubation with the primary antibody, mouse anti NG2/CSPG4 (clone NG 22161D7, gift from Roberto Perris, Centre for Molecular and Translational Oncology (COMT), University of Parma, Parma, Italy), overnight at 4 °C.

6. Rinse three times with 300 µL of PBS for 5 min at RT on a shaking platform.

7. Incubation with horse anti-mouse biotinylated antibody (Vector Laboratories, Inc., Burlingame, CA, USA; Cat No BA-2000) diluted 1:500 in BB subsequently revealed by

streptavidin-conjugated Alexa 488 (Thermo Fisher Scientific; Cat No S-32354) diluted 1:500 in BB for 45 min at RT in the dark.

8. Rinse three times with 300 µL of PBS for 5 min at RT on a shaking platform in the dark.

9. Postfixation in 4% PFA in PBS for 10 min at RT in the dark.

10. Rinse three times with 300 µL of PBS for 5 min at RT on a shaking platform in the dark.

11. Nuclear counterstaining by TOPRO-3 (diluted 1:10 K in PBS; Thermo Fisher Scientific) for 10 min at RT in the dark.

12. Transfer of sections for the adhesion on Vectabond™-treated slides (Vector Laboratories), coverslipping with Vectashield (Vector Laboratories), and sealing with nail varnish.

The slices were stored at 4 °C in the dark until performing the image acquisition.

For negative controls the primary antibody was omitted.

3.3 High-Resolution Confocal Imaging

The sections were examined under the Leica TCS SP5 confocal laser-scanning microscope (Leica Microsystems, Mannheim, Germany) setting the following parameters.

1. Laser activation at 488 nm and 633 nm wavelength for excitation of fluorophores in the green and red regions of the light spectrum, respectively, at appropriate laser power in order to avoid images oversaturation and samples fading.

2. Acquisition mode "xyz".

3. "Sequential frame acquisition".

4. "xy format".

5. "Acquisition speed" at 200 Hz (i.e., 200 lines/s).

6. "Image size" 1024 × 1024 pixels (246.03 mm × 246.03 mm) "pixel size" 240.50 nm × 240.50 nm (this image size setting increases the resolution for the analysis of cellular structures).

7. Section thickness 0.772 µm.

8. Value "line average" to 2 to reduce noise.

9. Fine-tuning of "gain and contrast" in the detection system to optimize signal intensity/noise ratio.

10. "z-stack" set plane "begin" and "end" of section for performing optical Z-stacks that span the whole volume of microvessels.

11. "z-step size" at the interval of 250 nm (this value depends on the thickness of the optical sections between 0.5 and 2.5 µm) through the z-axis for capturing z-stacks.

12. 63× oil immersion objective with a numerical aperture of 1.4 and optical resolution of 400 nm (immersion oil with a refractive index of 1.51).

3.4 Image Processing and 3D Reconstruction

The acquired z-stacks apply "process > tools > visualization > 3D projection > option maximum" of serial optical planes to obtain projection images using Leica LASAF software and Multicolour Package (Leica Microsystems).

To increase the resolution along the optical axis of the microscope and the signal-to-noise ratio of the acquired z-stacks, the deconvolution of the image data [30] is performed setting the following deconvolution parameters of Leica Confocal Software.

1. "Blind" deconvolution.

2. "Background removal" to determine the smallest intensity value in the image stack and detract it from the intensity values from all pixels in the image.

3. "Refractive index" set the value to 1.52.

4. "Set wavelength" select 520 nm for channel 1 and 587 nm for channel 2.

5. "Total iterations" set the value to 10.

6. "Rescaling intensity" and "resizing 16 bit" toggle off.

7. "Add new surface" to segment the vessel surface.

8. "Intensity," "Threshold" for each image to account for in background intensity and microvessel morphology.

9. Save the data.

4   Representative Results

The procedure of vibratome sectioning, followed by immunohistochemistry and confocal imaging, at the highest resolution, on thick sections on developing human brain allows for the visualization of a three-dimensional view of microvessels. The excellent preservation of cellular morphology shows the pericyte cell body and both the primary and secondary processes, to the fine ends, widely covering the microvessel wall (Fig. 1).

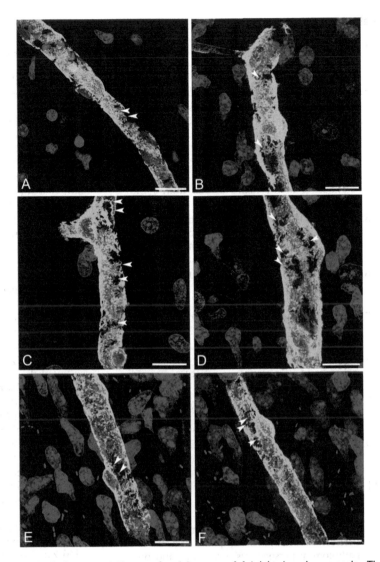

Fig. 1 (**a–f**) Representative confocal images of fetal brain microvessels. The deconvolution 3D on maximum intensity of projection images shows the structural characteristics of pericytes revealed by NG2 immunolabelling. At this resolution the individual pericyte primary and secondary processes (arrowheads) are clearly detectable. Scale bars (**a–f**) 10 μm

Acknowledgments

The authors gratefully acknowledge Prof. Roberto Perris for the generous NG 22161D7 antibody gift.

References

1. Armulik A, Abramsson A, Betsholtz C (2005) Endothelial/pericyte interactions. Circ Res 97 (6):512–523

2. Attwell D et al (2016) What is a pericyte? J Cereb Blood Flow Metab 36(2):451–455

3. Sweeney MD, Ayyadurai S, Zlokovic BV (2016) Pericytes of the neurovascular unit: key functions and signaling pathways. Nat Neurosci 19(6):771–783

4. McConnell HL et al (2017) The translational significance of the neurovascular unit. J Biol Chem 292(3):762–770

5. Virgintino D et al (2007) An intimate interplay between precocious, migrating pericytes and endothelial cells governs human fetal brain angiogenesis. Angiogenesis 10(1):35–45

6. Ribatti D, Nico B, Crivellato E (2011) The role of pericytes in angiogenesis. Int J Dev Biol 55 (3):261–268

7. Chiaverina G et al (2019) Dynamic interplay between pericytes and endothelial cells during sprouting angiogenesis. Cells 8(9)

8. Armulik A et al (2010) Pericytes regulate the blood-brain barrier. Nature 468 (7323):557–561

9. Berthiaume AA et al (2018) Pericyte structural remodeling in cerebrovascular health and homeostasis. Front Aging Neurosci 10:210

10. Dalkara T, Gursoy-Ozdemir Y, Yemisci M (2011) Brain microvascular pericytes in health and disease. Acta Neuropathol 122(1):1–9

11. Jansson D et al (2014) A role for human brain pericytes in neuroinflammation. J Neuroinflammation 11:104

12. Armulik A, Genove G, Betsholtz C (2011) Pericytes: developmental, physiological, and pathological perspectives, problems, and promises. Dev Cell 21(2):193–215

13. Blinder P et al (2013) The cortical angiome: an interconnected vascular network with noncolumnar patterns of blood flow. Nat Neurosci 16 (7):889–897

14. Zlokovic BV (2008) The blood-brain barrier in health and chronic neurodegenerative disorders. Neuron 57(2):178–201

15. Hartmann DA et al (2015) Pericyte structure and distribution in the cerebral cortex revealed by high-resolution imaging of transgenic mice. Neurophotonics 2(4):041402

16. Bonkowski D et al (2011) The CNS microvascular pericyte: pericyte-astrocyte crosstalk in the regulation of tissue survival. Fluids Barriers CNS 8(1):8

17. Dore-Duffy P, Cleary K (2011) Morphology and properties of pericytes. Methods Mol Biol 686:49–68

18. Smyth LCD et al (2018) Markers for human brain pericytes and smooth muscle cells. J Chem Neuroanat 92:48–60

19. Krueger M, Bechmann I (2010) CNS pericytes: concepts, misconceptions, and a way out. Glia 58(1):1–10

20. Winkler EA, Bell RD, Zlokovic BV (2010) Pericyte-specific expression of PDGF beta receptor in mouse models with normal and deficient PDGF beta receptor signaling. Mol Neurodegener 5:32

21. Nehls V, Drenckhahn D (1991) Heterogeneity of microvascular pericytes for smooth muscle type alpha-actin. J Cell Biol 113(1):147–154

22. Bandopadhyay R et al (2001) Contractile proteins in pericytes at the blood-brain and blood-retinal barriers. J Neurocytol 30(1):35–44

23. Chen J et al (2017) CD146 coordinates brain endothelial cell-pericyte communication for blood-brain barrier development. Proc Natl Acad Sci U S A 114(36):E7622–E7631

24. Alliot F et al (1999) Pericytes and periendothelial cells of brain parenchyma vessels co-express aminopeptidase N, aminopeptidase A, and nestin. J Neurosci Res 58(3):367–378

25. Ozerdem U et al (2001) NG2 proteoglycan is expressed exclusively by mural cells during vascular morphogenesis. Dev Dyn 222 (2):218–227

26. Stallcup WB (2002) The NG2 proteoglycan: past insights and future prospects. J Neurocytol 31(6-7):423–435

27. Tamburini E et al (2019) Structural deciphering of the NG2/CSPG4 proteoglycan multifunctionality. FASEB J 33(3):3112–3128

28. Girolamo F et al (2013) Diversified expression of NG2/CSPG4 isoforms in glioblastoma and human foetal brain identifies pericyte subsets. PLoS One 8(12):e84883

29. Winkler EA, Bell RD, Zlokovic BV (2011) Central nervous system pericytes in health and disease. Nat Neurosci 14(11):1398–1405

30. Biggs DS (2010) 3D deconvolution microscopy. Curr Protoc Cytom Chapter 12:Unit 12.19 1–Unit 12.1920

Quantification of Tumor Vasculature by Analysis of Amount and Spatial Dispersion of Caliber-Classified Vessels

Marco Righi, Marco Presta, and Arianna Giacomini

Abstract

This protocol focuses on the quantitative description of the angioarchitecture of experimental tumor xenografts. This semiautomatic analysis is carried out on functional vessels and microvessels acquired by confocal imaging and processed into progressively reconstructed angioarchitectures following a caliber-classification step. The protocol can be applied also to the quantification of pathological angioarchitectures other than tumor grafts as well as to the microvasculature of physiological tissue samples.

Key words Angiogenesis, Tumor vasculature, Spatial dispersion, ImageJ, Automatic analysis

1 Introduction

Quantification of vascular angioarchitectures represents a complex task. Indeed, vascular layouts differ according to the tissue examined [1] and microvessels show changing in their caliber and tortuosity [2] as well as differences in the frequency by which they stem from larger ducts [3].

So far, the mean vascularization index (MVI) has been widely utilized to describe a vascularized tissue [4, 5]. MVI represents the mean quantitative occurrence of vascular tissue in a specific spatial volume of the biological sample under analysis. In the past, many efforts were finalized to the description of vessels as an arrangement of "tubes" stemming from larger ducts. In this respect, their calibers, tortuosity as well as the frequency of vascular division [6–11] received intensive scrutiny but did not revealed any clue for a universal approach in the description and classification of an overall angioarchitecture. The possibility to provide a rationale description of microvascular angioarchitecture is utterly worsened when considering pathological samples, including the neoplastic tissue [12]. In tumors, the increased production of angiogenic factors changes the organized arrangements of capillaries in a confused

Domenico Ribatti (ed.), *Vascular Morphogenesis: Methods and Protocols*, Methods in Molecular Biology, vol. 2206, https://doi.org/10.1007/978-1-0716-0916-3_12, © Springer Science+Business Media, LLC, part of Springer Nature 2021

network of disorganized vessels [13] that cannot be described nor analyzed as single units.

In the past, in spite of significant efforts toward an unbiased description of tissue angioarchitectures, very little attention, if any, has been paid to the spatial vessel dispersion, a parameter of pivotal importance for a three-dimensional understanding of vessel distribution [14]. Intuitively, this parameter is crucial to describe whether vessels are well deployed and provide efficient delivery of oxygen and nutrients even at the extreme periphery of the tissue volume they supply. Despite its importance, this parameter is difficult to calculate because it requires the construction of an Euclidean distance map (EDM) [15] of the tissue volume, taking in account both vessels and volume boundaries. This problem was solved with the onset of the digital era by the development of image analysis programs and routines. In this context, the availability of the open access NIH application ImageJ [16] (named Fiji when preloaded with plugins of biological interest) [17] together with its scripting language, provided a simple, yet powerful, instrument to approach the problem of calculating vessel dispersion in three dimensions.

Assuming a binary stack representing the spatial vessel distribution in 3D, a possible approach consists in dilating the original signals in space in a sequence of fixed-radius steps following an approximated spherical dilation [18]. These steps would be carried cycle after cycle until the dilated signal occupies a defined amount of volume (e.g., 95%), thus providing an index (dubbed nHv 95%) after its normalization based on the amount of input voxels. The total number of cycles needed to obtain this result is a direct consequence of the initial signal dispersion in the volume, with fewer cycles indicating a more uniform and pervasive dispersion of signal. Drawbacks are to be found in the time needed to complete the analysis, which is lengthy and resource consuming even for today computers. Alternative, faster approaches are possible, even if based on different axioma and therefore differently approximated [19]. The protocol herewith described will utilize an approximated spherical dilation that we implemented as a rhombicuboctahedral dilation scheme.

In addition to dispersion, an angioarchitecture is characterized by the amount and calibers of the vessels which concur to its layout. This signature is characteristic of a given angioarchitecture and changing in the proportions of smaller versus larger capillaries alter the overall amount of fresh blood available to the tissue in a given period of time [20]. Unfortunately, the picture is blurred by the observation that these parameters are not linked by straightforward relationships. In this respect, our initial attempt to correlate the amount of caliber-classified vessels to the extent of their dispersion failed [21]. However, this goal was achieved when we addressed the following question: given the initial setup of the largest vessels, how such an arrangement would evolve in terms of signal amount and signal dispersion upon consideration of vessels with progressively smaller calibers? It turned out that this evolution is often near-linear and can be summarized by linear regression analysis in tumor samples [22] as well as in more physiological angioarchitectures [23].

In the present work we describe an experimental protocol to analyze the functional microvascular network of human tumor xenografts in mice. The protocol, applied to image analysis of sulfobiotin stained tumor sections, shows that amounts, spatial dispersion and spatial relationships of adjacent classes of caliber-filtered microvessels provide a near-linear graphical "fingerprint" of tumor vascularization in which position, slope and axial projections of this graphical outcome reflect biological features and summarize the properties of the tumor microangioarchitecture.

2   Materials

1. Tumor functional vasculature is stained by in vivo biotinylation using sulfosuccinimidyl-6-(biotinamido) hexanoate (sulfo-NHS-LC-biotin, Thermo Fisher Scientific, Rockford, IL, USA). Biotinylated vessels of tumor xenografts are then revealed ex vivo by staining with Alexa Fluor 488-conjugated streptavidin (Invitrogen, Carlsbad, CA, USA). Reagents needed:

   (a) Sulfo-NHS-LC-biotin (5 mg/mL) in polygeline.

   (b) Neutralization solution: Tris/polygeline (50 mM).

   (c) Fixative: 4% paraformaldehyde.

   (d) Embedding: 6% agarose in water.

   (e) Blocking solution: 2% BSA in PBS containing 0.2% Triton X-100.

   (f) Alexa Fluor 488-conjugated streptavidin diluted 1:200 in PBS containing 0.2% BSA and 0.1% Triton X-100.

   (g) Rinsing solution: PBS containing 0.05% Tween 20.

   (h) Vectashield mounting medium (Vector Laboratories).

2. Images are acquired with a confocal microscope (MRC-1024 Bio-Rad, Bio-Rad Laboratories Inc., UK) equipped with a 40× oil immersion objective (NA 1.0) at lambda 488. Pixel dimension is calculated by the accompanying software (LaserSharp, Bio-Rad Laboratories Inc., UK).

3. The 64 bit ImageJ Java application v. 1.50e run under Java SDK 1.8.0_66 is used throughout the analyses. This application, preloaded with a set of biological plugins, is available also under the name Fiji [17] and retains the ability to run the

different scripts detailed in Subheading 3, provided the presence of an appropriate and updated Java Standard Development Kit (SDK).

ImageJ versions earlier than v. 1.50 are not recommended.

4. Some of the macros developed for this analysis are dependent on the TransfomJ sets of plugins developed by J. Meijering [24] and freely available at ImageScience.org (https://imagescience.org/meijering/software/transformj/). Once downloaded, the recovered .jar files are to be moved to the "ImageJ/plugins/jar" subfolder (or to any location indicated by the accompanying instructions) and the ImageJ program must be restarted.

5. Scripts are available at the GitHub repository https://github.com searching for "nHv95/Vessel_analysis". Scripts with a ".txt" extension are identified by name and their code is totally accessible. The scripts detailed in this protocol are written in the ImageJ macro language [25]. The plugin "Spatial_dispersion.class" used in this sample analysis can be also obtained compiling the "Spatial_Dispersion.java" text provided at the same repository (*see* **Note 1**) and moving the resulting ".class" file in one subfolder of the plugins folder of the ImageJ application. Code is not directly accessible from compiled plugin, but can be found in the .java text ready to be compiled. Plugin is written taking advantage of version 1.71 of a specific tutorial authored by W. Bailer [26]. Rationale behind plugin code is summarized at the end of this protocol.

6. Eventual statistical analyses can be carried out using the statistical package Prism from GraphPad Software run on an Apple Macintosh Pro computer.

3   Methods

3.1   Vessel Staining

To analyze tumor vasculature, proteins expressed on the luminal surface of vascular endothelium are in vivo biotinylated by intravenous injection of sulfo-NHS-LC-biotin. This compound is impaired in diffusion through biologic membranes and efficiently allows for the in vivo biotinylation of luminal vascular endothelial proteins [27]. Briefly:

1. subcutaneously tumor-bearing animals are intravenously injected with 0.2 mL of sulfo-NHS-LC-biotin (5 mg/mL) in polygeline.

2. After 5 min, mice are intravenously injected with 1 mL of Tris/polygeline (50 mM) to neutralize the circulating reagent [28].

3. Biotinylated tumors are then excised, fixed in 4% paraformaldehyde overnight at 4 °C and embedded in 6% agarose.

4. Agarose-embedded tumors are cut at a thickness of 100 μm using a vibratome.

5. Free-floating vibratome tumor sections are blocked with 2% BSA in PBS containing 0.2% Triton X-100 for 1 h at room temperature.

6. After blocking, sections are incubated overnight at 4 °C with Alexa Fluor 488-conjugated streptavidin diluted 1:200 in PBS containing 0.2% BSA and 0.1% Triton X-100. Then sections are rinsed 4 times for 30 min in PBS containing 0.05% tween20.

7. Finally, sections are mounted with Vectashield mounting medium in order to perform confocal microscopy analysis.

3.2 Image Acquisition

Z-series of vascular fields from 100 μm-thick sections are acquired using a standard confocal microscope equipped with a 40× oil immersion objective (NA 1.0) at lambda 488. Pixel dimension is calculated by the accompanying LaserSharpsoftware and in the case of this analysis resulted in a size of 0.54 μm. In the vertical axis, slices were 1 μm apart.

3.3 Images to Stack

1. Color images must be converted to grayscale images (*see* **Note 2**). To this goal it is better to use the command "Image > Color > Split Channels" on each image. Images from the appropriate color channel, representative of vascular staining, are to be grouped together whereas other images can be discarded.

2. Group all the 8-bit grayscale TIFF images relative to a same confocal stack into a single folder taking care that their ordinate listing from top to bottom represents their top to bottom orientation in the acquisition sequence (*see* **Note 3**).

3. Using the ImageJ application, select "File > Import > Image sequence" indicating the position of the source folder.

4. Check that all parameters that will appear in the Sequence Options menu are correct. In particular: the total number of images, the starting image (usually 1), the increment to the next image [1] and the scale of the final images (100%). If required, filter images to be inserted by name or regex patterns.

5. After visual inspection of the stack for correctness, save the novel stack in its intended folder.

3.4 Making Stack Isotropic

1. Using the ImageJ application, open the stack to be made isotropic by "File > Open" by selecting the appropriate file.

2. Select "Image > Properties" to verify for voxel dimensions in order to control for appropriate image resolution. In the pop-up panel control that parameters Channels and Frames are equal to 1, whereas Slices equal the number of the slices

of the stack. Check the unit of length for correctness as well as Pixel width and height according to original image resolution. Control that Voxel depth correspond to the mean distance among scans used by the confocal microscope at the moment of image acquisition (*see* **Note 4**). Frame interval is to be set to 0 s and Origin to 0,0 value.

3. In presence of the open image stack select command "Plugins > Tools > Make isotropic". Then save the resulting stack in a separate folder (*see* **Note 5**).

3.5 Contrast Enhancement and Image Thresholding

1. In order to increase the chances of obtaining representative binary images, it is mandatory to optimize grayscale image contrast. This operation is carried out in the presence of the open isotropic stack calling the ImageJ command "Process > Enhance Contrast" and processing the resulting pop-up choices.

2. Call the ImageJ command "Process > Enhance Contrast".

3. Indicate the maximal percent of saturated voxels in the resulting stack (*see* **Note 6**).

   (a) Select whether you wish to equalize stack histogram (rule of thumb: NO, *see* **Note 7**).

   (b) Select whether you wish to process all slices (YES).

   (c) Select whether you want to use stack histogram (YES, *see* **Note 8**).

   (d) Click OK and save the resulting contrasted stack (Fig. 1, Panel a) in its proper folder.

4. Image Thresholding is carried out according to one of the already available ImageJ filters. The choice of the correct filter is rather arbitrary, being dependent on image acquisition and on the overall image story following image acquisition. The Authors tend to concentrate on two different methods: the Default method which is somehow permissive, artefactually increasing presumed vascular signals, and the Renji Entropy method which is somehow restrictive, underestimating vascularization (*see* **Note 9**).

5. Using the ImageJ application, open the grayscale image stack to be thresholded and select "Process > Binary > Make Binary". From the pop-up control panel select the appropriate thresholding method and then apply the transformation.

6. Save the resulting binary image stack in a different folder from its grayscale precursor (Fig. 1, Panel b).

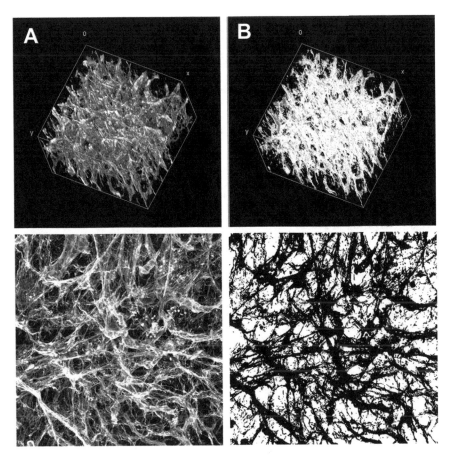

Fig. 1 Representations of grayscale and binary vascular angioarchitectures. Panel (**a**) Top: image rendering of a contrast-normalized, isotropic, grayscale stack representing the confocal signals recovered after staining of functional vessel walls of an experimental tumor with Alexa Fluor-488 streptavidin following sulfo-NHS-LC-biotin priming. Bottom: a z-projected image of the same graystack stack highlights the vascular complexity of the experimental tumor vasculature. Panel (**b**) Top: image rendering of the same volume as in Panel A, after thresholding to obtain a binary stack. Bottom: z-projection of the binary stack. Images underscore the amount and quality of binary signals as well as the vascular representativeness of the binary volume

3.6 Vessel Fill-Up [macro: 3D_Close&Fill. txt]

1. Vessel calibers are to be calculated according to the lowest vascular cross-section drawn in the three Euclidean planes. The area of these cross-sections is obtained summing up all the appropriate contiguous binary voxels related to signal. Given that sulfobiotin-mediated staining identify only vascular walls, vessels might show pervious lumens resolving in unwanted holes in their cross-sections depending on their orientation.

   To bypass this problem, in this step we start a parallel workflow with duplicated images in order to obtain caliber masks that will be used at Subheading 3.8 to assign correct calibers to original signal voxels. Here, we will be adding extra signals to try to close incomplete vessel walls and to fill void

Planning of the analysis

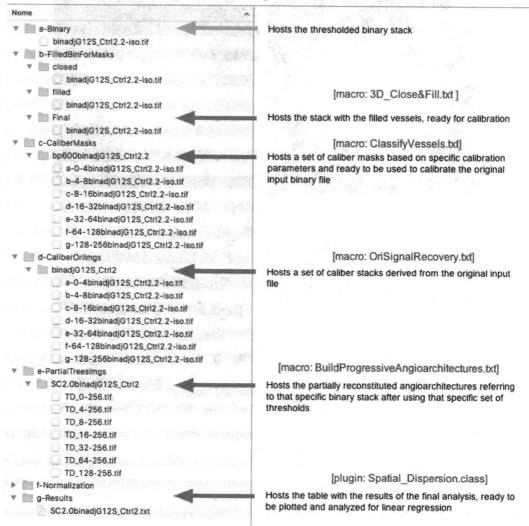

Nome	
▼ 📁 a-Binary	Hosts the thresholded binary stack
binadjG12S_Ctrl2.2-iso.tif	
▼ 📁 b-FilledBinForMasks	
▼ 📁 closed	
binadjG12S_Ctrl2.2-iso.tif	
▼ 📁 filled	[macro: 3D_Close&Fill.txt]
binadjG12S_Ctrl2.2-iso.tif	
▼ 📁 Final	Hosts the stack with the filled vessels, ready for calibration
binadjG12S_Ctrl2.2-iso.tif	
▼ 📁 c-CaliberMasks	[macro: ClassifyVessels.txt]
▼ 📁 bp600binadjG12S_Ctrl2.2	Hosts a set of caliber masks based on specific calibration
a-0-4binadjG12S_Ctrl2.2-iso.tif	parameters and ready to be used to calibrate the original
b-4-8binadjG12S_Ctrl2.2-iso.tif	input binary file
c-8-16binadjG12S_Ctrl2.2-iso.tif	
d-16-32binadjG12S_Ctrl2.2-iso.tif	
e-32-64binadjG12S_Ctrl2.2-iso.tif	
f-64-128binadjG12S_Ctrl2.2-iso.tif	
g-128-256binadjG12S_Ctrl2.2-iso.tif	
▼ 📁 d-CaliberOriImgs	[macro: OriSignalRecovery.txt]
▼ 📁 binadjG12S_Ctrl2	Hosts a set of caliber stacks derived from the original input
a-0-4binadjG12S_Ctrl2.2-iso.tif	file
b-4-8binadjG12S_Ctrl2.2-iso.tif	
c-8-16binadjG12S_Ctrl2.2-iso.tif	
d-16-32binadjG12S_Ctrl2.2-iso.tif	
e-32-64binadjG12S_Ctrl2.2-iso.tif	
f-64-128binadjG12S_Ctrl2.2-iso.tif	
g-128-256binadjG12S_Ctrl2.2-iso.tif	
▼ 📁 e-PartialTreesImgs	[macro: BuildProgressiveAngioarchitectures.txt]
▼ 📁 SC2.0binadjG12S_Ctrl2	Hosts the partially reconstituted angioarchitectures referring
TD_0-256.tif	to that specific binary stack after using that specific set of
TD_4-256.tif	thresholds
TD_8-256.tif	
TD_16-256.tif	
TD_32-256.tif	
TD_64-256.tif	
TD_128-256.tif	
▶ 📁 f-Normalization	[plugin: Spatial_Dispersion.class]
▼ 📁 g-Results	Hosts the table with the results of the final analysis, ready to
SC2.0binadjG12S_Ctrl2.txt	be plotted and analyzed for linear regression

Fig. 2 Analytical flowthrough. The left portion of the figure groups folders and resulting volumes deriving from the progression of the analysis throughout the different steps. The right portion explains the role of selected folders or subfolders and reports which macro (blue names) should be used to progress to the next analytical step. Starting from the stack hosted in the initial folder indicated by the green arrow, the analysis run in a top-down progression with each red arrow marking where to find the results given by the macro listed immediately upstream. Not less than five automatic routines are necessary to obtain the final result table. However, presence of intermediate results allows for their analysis, in case of unexpected results or behaviors

vessels before determining their cross-sections. These signals will be removed after vessel calibration. Figure 2 shows the flow through of the elaboration starting with the binary stack.

2. Using the ImageJ application, select "Plugins > Macro > Run" and then indicate macro 3D_Close&Fill.txt The macro will ask for a source folder containing all the stacks to be filled; folder

that is named a-Binary in Fig. 2. Then, the macro will ask for a folder were to save the resulting filled stack (b-FilledBinForMasks).

3. During the elaboration the macro identifies and fill holes (areas of white signal surrounded by contiguous black voxels) in each slice according to two synergistic approaches (*see* **Note 10**). This step is automatically repeated slice after slice and according to the three Euclidean orientations of the volume. The resulting stack will be equidimensional with the input volume but will contain a larger amount of black voxels that will be discarded after calibration. The output folder will list three service subfolders, namely, "closed," "filled," and "Final" (Fig. 2). This last subfolder will list the filled volumes ready for calibration.

3.7 Bandpass Caliber Masks [macro: ClassifyVessels.txt]

1. A calibration step implies a defined set of dimensional thresholds. The script used in this step allows the user to choose among sets of predefined thresholds, filtering one or more stacks sequentially. Given that resolution can vary among analyses, all values are expressed in square pixels (or voxels). Treatment of each input stack will generate a folder listing seven stacks with the desired bandpass voxels. Voxels belonging to particles larger than the largest threshold will not be considered and will be discarded.

    Thresholds are predefined according to a couple of rules:

    (a) The lowest threshold is called "offset" and can be chosen by the user among five dimensions (in square pixels) ranging from 4 to 64. It identifies the lowest class of calibers, those showing a cross-section comprised between "offset" and 0 square pixels (*see* **Note 11**).

    (b) All other thresholds are sequential multiples of "offset," rounded to integers. To set their values the user can choose a multiplier among six values ranging from 1.5 to 2.0 in 0.1 increments. The chosen value will be multiplied with "offset" and, separately, with the numbers 2, 3, 4, 5, 6 to give the selected thresholds.

    As an example, thresholds used for tumor vessels in Righi et al. [21], were defined with an offset of 4 and 2.0 as multiplier, and here we will behave accordingly. Taking in account the selectable choices and the resolution of the original image stack, it is thus possible to tailor the dimensions of the considered vessels in order to restrict or expand the analysis to sizes really present in the tissue context (*see* **Note 11**).

2. Using the ImageJ application, select "Plugins > Macro > Run" and then indicate macro ClassifyVessels.txt The macro will present a dialog interface in which the user should define:

(a) "offset," choosing from a multiple-choice menu with pixel values 4, 8, 16, 32, 64.

(b) the multiplier, choosing from a multiple-choice menu with values 1.5, 1.6, 1.7, 1.8, 1.9, 2.0.

3. Then the macro asks for a folder grouping all the input volumes (subfolder "Final" in Fig. 2), and for a folder where it will save all the output stacks (c-CaliberMasks). It is mandatory that the input volume has a square basis and an even number of slices. If the slice number of the stack is odd, it will be automatically converted to even by deletion of the first slice. If the basis of the volume is not a square (e.g. a rectangle), the macro will abort and exit (*see* **Note 12**).

4. Actually, output is represented by a result folder for each processed stack. The name of the subfolder begins with the "bp" acronym and groups the seven bandpass stacks reporting the locations of the voxels belonging to particles with a specific caliber (Fig. 2). The size of the caliber is comprised between the two values, in pixels, indicated in the name of the file and defined mathematically after choosing the initial parameters.

3.8   Classification of Original Binary Voxels in Caliber Classes [macro: OriSignalRecovery.txt]

1. This step is carried out to classify the original binary signals in dimensional classes according to the caliber masks obtained in Subheading 3.7, **step 4**. The macro needs two input folders and can work only with strictly equal numbers of masks and angioarchitectures. It also verifies the correspondence of the name of the original stack with the name of the caliber masks and aborts in case of mismatch.

2. Removal of service files. Only the folder beginning with the "bp" acronym and obtained in Subheading 3.7, **step 4** can be used in successive steps. In previous versions of the elaboration, an additional service folder should have been manually removed for each sample volume before launching the next routine. This step has now been implemented into the latest version of macro ClassifyVessels.txt and is no more necessary, but can be mandatory using previous version of the same macro.

3. Using the ImageJ application, select "Plugins > Macro > Run" and then indicate macro OriSignalRecovery.txt As a first message, the macro asks to select the folder with the original binary volume, namely that saved at Subheading 3.5, **step 6** and hosted in a-Binary in Fig. 2. Then it asks to select the folder containing the folders with the appropriate caliber masks (*see* **Note 13**) and which is named c-CaliberMasks in our example. Finally, it asks for the folder in which to save the results. In our planning, it is called d-CaliberOriImgs.

4. Provided that the number of binary input volumes and mask subfolders is equal, and that the name of the input volume correspond to the name of the mask subfolder, the macro executes. Results are represented by a folder grouping seven binary stacks reporting the positions of voxels belonging to specific caliber classes and obtained from the distribution of the original vascular signals. Dimensional classes are identified by two values (in pixels) stated in the filename of the stack.

3.9 Assembly of Progressively Reconstructed Angioarchitectures (PRA) [macro: BuildProgressive-Angioarchitectures. txt]

1. This step combines specific groups of caliber-classified voxels in order to prepare a set of seven PRA, where the first reports only the voxels belonging to the largest vessels considered. Information from this stack are then integrated with those from vessels of immediately lower caliber to obtain the second analytical sample. The process is carried on recursively until the final, seventh analytical sample is representing the fully reconstituted angioarchitecture, presenting signals from all the considered vessels down to the smallest ones (Fig. 3).

2. In order to run this script, passband images SHOULD be organized in a 2-level nested folder and named in order to be listed from lower to higher calibers (*see* **Note 14**).

3. Using the ImageJ application, select "Plugins > Macro > Run" and then indicate macro Build Progressive Angioarchitectures. txt. As a first message, the macro asks to define which multiplier was used at Subheading 3.7, **step 2** when calibrating vessels (2.0 in our case), in order to report it in the results. Then the macro asks to select the folder grouping subfolders with caliber-classified   images   ready   to   be   combined (d-CaliberOriImgs). Finally, it asks for the folder in which to save the results.

4. Upon completion, results are represented by a folder characterized by an initial "SC" acronym followed by the value of the multiplier and by the name of the sample (*see* **Note 15**). This folder groups seven binary stacks characterized by an initial "TD" acronym followed by the min-max values in pixels of the cross-sections of the vessels grouped in the stack (Fig. 4). These vessels are now ready for the analysis, except that they lack normalization that is mandatory when comparing two or more samples.

3.10 Normalization [macro: NormalizingVolume. txt]

1. When comparing the signal dispersion of several sets of vascular angioarchitectures, it is necessary to cope with different amounts of input signals. Presence of different amounts of signals will provide a different value for spatial dispersion even in the presence of identical distributions. A possibility to "normalize" the input voxels of each sample relies on the initial dilation of each angioarchitecture until attaining the largest

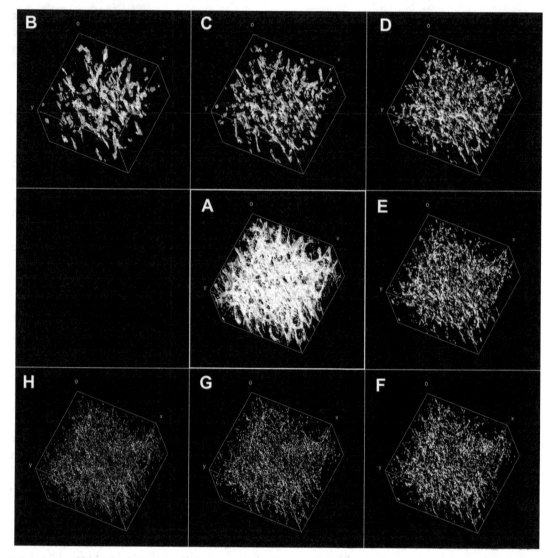

Fig. 3 Decomposition of filled vascular masks according to selected classes of calibers. The central rendering (**a**), representative of the filled vascular stack, is shown rounded by the renderings of the stacks (**b–h**) representing signals belonging to different classes of calibers after decomposition of stack A. (**b**) showing signals with cross-sections from 75 to 37 μm; (**c**) from 37 to 19 μm; (**d**) from 19 to 9.5 μm; (**e**) from 9.5 to 4.7 μm; (**f**) from 4.7 to 2.4 μm; (**g**) from 2.4 to 1.2 μm; (**h**) from 1.2 to 0.0 μm

amount of signal observed among all the samples to be compared. It is thus mandatory to identify the sample with the largest amount of signal voxels.

2. In our setup, using the ImageJ application, select "Plugins > Macro > Run" and then indicate macro Normalizing-Volume.txt. The macro will perform the assessment of the percent volume occupied by signal (black voxels) in a binary volume. The macro will ask to select the folder containing

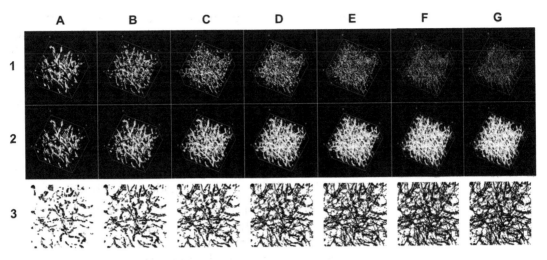

Fig. 4 (**a–g**) Distribution of signals in different classes of calibers and partially reconstructed angioarchitectures. The Figure is organized by rows and columns. Top row (**1**): renderings of vascular signals classified into seven different classes of calibers from largest (**a**) to smallest (**g**). Middle row (**2**): Partially Reconstructed Angioarchitectures where an image is representing the sum of the image on its left with the corresponding image in the top row. PRAs evolves from largest signals (**a**) to totally reconstructed angioarchitecture (**g**). Bottom row (**3**): similar to row 2, but images represent *z*-projections of the corresponding stacks instead of renderings. Renderings (**1a**) and (**2a**) are equals as they are built from the same volume of large vessels

subfolders, each representative of a different sample. Each subfolder will group all the PRAs relative to that sample. Then the macro will ask to choose the position in which to save the resulting files (a file for each sample, not for each PRA).

3. Results will list the different values relative to the percent volume occupied by signal voxels reporting at the end the name of the volume with the largest value observed among those of this sample (*see* **Note 16**). Upon identification of the sample with the largest signal, normalization is carried out by hand, copying the selected volume and pasting it in each folder harboring the set of PRAs for a given sample (*see* **Note 17**).

4. The normalization step is performed manually and can be sensibly improved in terms of automation, both in the identification of the volume with the largest signal as well as in the pasting of the resulting volume. Further versions of the script might be modified accordingly.

3.11 Analysis of PRAs by Signal Amount and Spatial Dispersion [plugin: Spatial_Dispersion. class]

1. This point implements the final step and it is finalized to quantify both signal amount and normalized spatial dispersion for each PRA of a given sample. For speed reasons, this analysis is carried out using a dedicated plugin whose rationale is explained in details in the appropriate chapter of this protocol.

2. Using the ImageJ application, select "Plugins > ", then the folder where you moved the compiled plugin and then "Spatial Dispersion". The plugin will ask for a folder grouping all the subfolders in analysis, each grouping the PRAs for a specific sample together with the normalizing volume, if any. Then the plugin will ask for the location in which to save the result files. Upon execution the plugin will show a progress bar to monitor the advancement of the analysis (*see* **Note 18**).

3. At the end of the elaboration, results are represented by a set of .txt files named after the names of the analyzed samples/subfolders. Each file reports an (a) header and a (b) table.

(a) In the header the macro reports the location of the folder chosen for analysis and the starting time of execution. Then it reports the percent volume (Vol%) of normalization and the name of the normalizing stack. These data are a precaution against the possibility of errors or omissions in normalization.

(b) The table is organized in nine rows and eight columns with values separated by tab commands.

After a row for captions, any of the following rows is dedicated to a PRA, apart an additional row for the normalizing volume. As for the columns they reports left to right (Fig. 5):

(a) the name of the volume (PRA or normalizer).

(b) the Vol% occupied by signal (black voxels).

(c) the raw Hv 90% index (i.e. the total number of dilation cycles needed to occupy 90% of the volume).

(d) the raw Hv 95% index.

(e) the raw Hv 99% index.

(f) the normalized nHv 90% index (the Hv 90% index less the number of cycles needed to reach a starting Vol% equal to normalizer).

(g) the normalized nHv 95% index.

(h) the normalized nHv 99% index.

The table can be imported into Excel, Numbers or Open-Office spreadsheet applications, minding the number format which follows the USA style. Unfortunately, the plugins does not process volumes in the desired order (Fig. 5, Panel a), because ImageJ samples names character by character and not by overall value. Although this behavior can be corrected when creating PRAs, this has not yet been implemented and rows must be ordered by hand (Fig. 5, Panel b) to prepare for linear regression.

● ● ●   📄 SC2.0binadjG12S_Ctrl2.2-iso.txt — Modificato ⌄

Calculating normalized Hv indexes in nested folders.
Plugin Ultra_nHvIndex Version: 1.0 del 30-04-2012   **A**

Analyzed folder of folders: /Users/ghirigoro/Documents/Lavoro/Progetti/c-SezCumulate-AntiAngioIndex/c-DatiXsezioniCumulate/
p-Ctrl4analysis/e-PartialTreesImgs/
At time: 5-nov-2019 15.16.53

Vol % di normalizzazione: 2.663928882495777   relativo al volume: TD_0-256.tif

Immagine:	Vol %:	Hv90%:	Hv95%:	Hv99%:	nHv90%:	nHv95%:	nHv99%:
TD_0-256.tif	2.663929	14.622073	17.281843	22.615501	14.622073	17.281843	22.615501
TD_128-256.tif	0.75059223	40.83803	48.384968	62.554863	39.28165	46.828587	60.998478
TD_16-256.tif	1.9580408	21.582438	25.656183	35.902573	21.364674	25.438417	35.684807
TD_32-256.tif	1.6639668	25.296495	30.248035	44.317272	24.91084	29.86238	43.931618
TD_4-256.tif	2.256014	17.062984	19.925007	25.753311	16.96835	19.830372	25.658676
TD_64-256.tif	1.2818909	30.701677	36.743507	51.886974	29.981585	36.023415	51.16688
TD_8-256.tif	2.147254	18.994007	22.315626	29.418665	18.858723	22.180342	29.283379

Fine elaborazione alle: 15.23.56

B

Immagine:	Vol %:	Hv90%:	Hv95%:	Hv99%:	nHv90%:	nHv95%:	nHv99%:
TD_128-256.tif	0.75059223	40.83803	48.384968	62.554863	39.28165	46.828587	60.998478
TD_64-256.tif	1.2818909	30.701677	36.743507	51.886974	29.981585	36.023415	51.16688
TD_32-256.tif	1.6639668	25.296495	30.248035	44.317272	24.91084	29.86238	43.931618
TD_16-256.tif	1.9580408	21.582438	25.656183	35.902573	21.364674	25.438417	35.684807
TD_8-256.tif	2.147254	18.994007	22.315626	29.418665	18.858723	22.180342	29.283379
TD_4-256.tif	2.256014	17.062984	19.925007	25.753311	16.96835	19.830372	25.658676
TD_0-256.tif	2.663929	14.622073	17.281843	22.615501	14.622073	17.281843	22.615501

Fine elaborazione alle: 15.23.56

Fig. 5 Result table. Result tables after the final analysis of seven PRAs from our sample. Table (**a**) is the raw output of the plugin which reports the normalizing volume used (the fully reconstituted angioarchitecture). Table (**b**) presents data as they should appear, reorganized in serial order

3.12 Linear Regression and Important Descriptive Parameters

1. The scope of data elaboration is to extract relationships among PRAs of a same sample and to identify a reduced set of parameters able to describe the overall angioarchitecture analyzed.

2. In your preferred spreadsheet application, remove or marginalize normalizing information, if any, by moving them aside or out of the main result table. Then, order result table rows by hand to obtain a seven rows ramp table of increasing Vol% (or decreasing sizes in names) as in Fig. 5, Panel b.

3. Evaluate linear relationships between Vol% and your preferred normalized H index (e.g. nHv 95%) by linear regression (LR) over the seven values of the results table. Estimate linearity by checking the R^2 value (*see* **Note 19**).

4. Plot the results from the result table and from **step 3** in a XY graph showing Vol% on the X-axis and nHv values (e.g. nHv 95%) on the vertical axis. Plot both the seven result points from PRAs and the derived LR line to look how LR agrees with experimental points (Fig. 6, Panel a). This plot can be related to a rendering of the initial vascular layout in order to confirm the capacity of the LR line in summarizing the original microangioarchitecture (*see* Subheading 3.15).

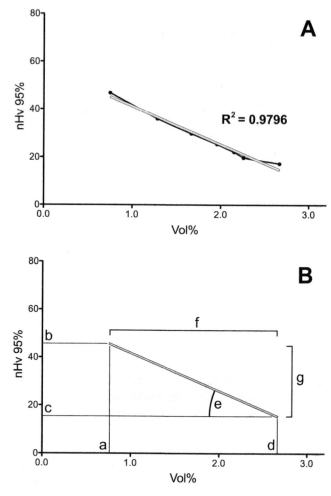

Fig. 6 Graphic analysis. Graphical plot of the data reported in Fig. 5. Panel (**a**): Plot of amount and spatial dispersion data from result table (black dots) connected by a black line. The yellow line is the linear regression obtained from data, and the R^2 value reported highlights the goodness of fit. Panel (**b**): Listing of descriptive parameters that can be derived from LR and map line position, slope, length and height. For the biological meaning of the descriptive parameters refer to Subheading 3.14 in the main text

5. Using the max and min Vol% values from the result table and the LR equation estimate the coordinates of the extremes of the LR line (a, b and c, d in Fig. 6, Panel b) as well as its slope (e). Using these data, derive the descriptive parameters of the line: length (f) and height (g) between extremes.

   X and Y positions of both extremes, slope, length and height of the line are the set of seven parameters that describe the micro-angioarchitecture just analyzed and that can be used for statistical analysis.

3.13 Biological Meaning of Descriptive Parameters

1. Each of the seven parameters that can be derived from the LR obtained using data from the seven PRA has its own biological meaning.

   In this respect, parameters from the left end of the line refer to the layout of signals from the largest vessels considered in the analysis. The X value reflects the amount of signals of this class of vessels. Similarly, the Y value concerns dispersion of these same signals, with large values indicating poor dispersion and high clusterization.

   Data from the right end of the LR are related to the overall angioarchitecture analyzed; from the largest vessels to the smallest capillaries. Here, again the X value indicates the total amount of signal and the Y value the maximal signal dispersion achieved by that vascular layout.

   The length of the LR declares the amount of signal from vessels smaller than the largest ones. At a difference, its height refers to the improvement in signal dispersion obtained when considering all the vessels smaller than the largest class. Finally, slope of the LR conveys a more elusive parameter: the improvement in signal dispersion as vessels become smaller and smaller. This parameter states the contribute of smaller vessels to cargo transport, in addition to delivery. A higher absolute value indicates an inefficient angioarchitecture where low-flow capillaries are deputed to distribute cargo in the volume as an additional task to cargo release.

   Comparison with a reference angioarchitecture allows mapping the LR for a test angioarchitecture in a specific region of the resulting plot. This mapping contributes to betray the anti- or proangiogenic status of the tissues surrounding the test angioarchitecture. It is not possible to quantitate this status in absolute terms as its value acquire meaning only when compared with a reference angioarchitecture. In this respect, and considering the different aspects involved, it might be difficult to evaluate subtle differences between two sets of results. This task is, however, easy when comparing a steep LR that lie in the top-left quadrant of an amount/spatial dispersion plot with a flat line confined in the bottom-right quadrant of the same graph. In these conditions, the steepest curve would show a poor angiogenic status with respect to the second line, presenting lower amounts of signals inefficiently dispersed in space.

3.14 Visual Workflow Control by Volume Viewing (3D Renderings)

1. Visual renderings of 3D volumes are very helpful in order to verify the original vascular layout and the actual vessels analyzed. In this respect, the ImageJ application provide a default plugin that the Authors find very useful (Fig. 7, Panel a).

2. Open the contrast-enhanced, isotropic grayscale stack saved after Subheading 3.5, **step 4** just before binarization.

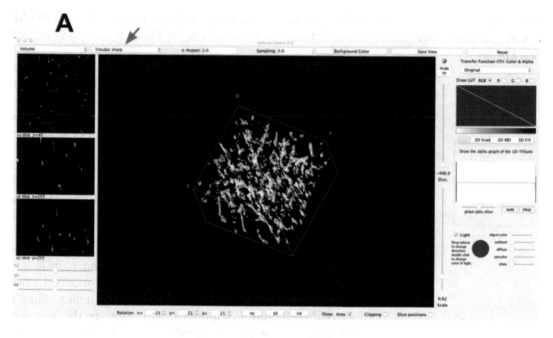

A

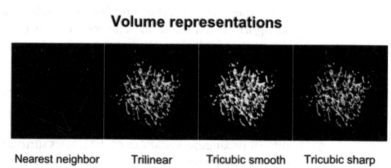

B   Volume representations

Nearest neighbor   Trilinear   Tricubic smooth   Tricubic sharp

Fig. 7 Graphical interface to volume viewer. Panel (**a**) Overall view of the interface; the red arrow points to the choice of the rendering method. Panel (**b**) Influence of rendering methods on volume representation. The panel reports results observed with the same volume but using different rendering methods

3. Call the ImageJ Volume Viewer plugin at "Plugins > 3D > Volume Viewer". To set preferences select, in the top line of commands "Volume" and "Tricubic sharp" from the two multiple-choice menu on the left, ignoring the other commands for the moment. Difference in choosing among the available renderings is shown in Fig. 7, Panel b. At this point, near the right border of the graphical interface, set the Distance and Scale values. For a $512 \times 512 \times 100$ vx stack the Authors advice for lowest Distance (-728.0) and unitary Scale (1.0). Finally, in the bottom line set rotations to your preferred values. We used the $X = -21$, $Y = 21$, $Z = 21$ values but obviously other combinations can be chosen.

To improve rendering, you can increase sampling value (top line) to 5 or more, but the time to elaborate the final image will be increased accordingly. After several trials, a value of 5 appears however quite satisfactory. At any time, you can obtain an image of the actual rendering clicking on the Save View button in the top-right corner.

3.15 Identification of Discarded Vascular Structures and Signals

1. As a final control, Volume Viewer can be used to visualize the eventual vessels or structures too large to be analyzed. This step will highlight also differences in the original binary image and the reconstructed final angioarchitecture used for analysis.

2. Open the original binary file. Then, open the fully reconstructed angioarchitecture that was created at Subheading 3.9, **step 4** and is named TD_0-xxx.tif, with xxx being the maximal threshold used for vessel calibration (256 in our example). This last image stack represent the largest PRA ready for analysis and differentiates from the binary sample only because it lacks voxels belonging to vessels with calibers larger than the maximal threshold used.

3. Using ImageJ image calculator "Process > Image Calculator … " subtract the fully reconstructed angioarchitecture from the original binary sample, asking to process all the images of the stack. The result will appear as a binary stack with an inverted LUT that can be directly used for rendering (Fig. 8).

3.16 Analysis Scale-Up

1. The described routines are ready to analyze more than a sample. In this respect, it makes sense to indicate a folder grouping ALL the samples to be analyzed as the source folder for all macro. All the resulting images will be grouped in the target folder already ordered to be ready for analysis with the following routine.

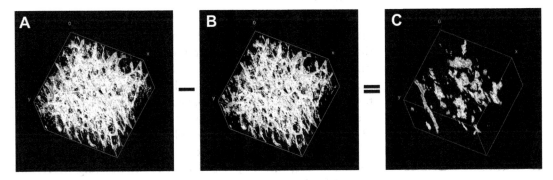

Fig. 8 Structures excluded from analysis. Signals belonging to very large structures are not classified in caliber classes and therefore are excluded from analysis. The figure presents the renderings of the original binary volume (**a**) from which we subtract (**b**), the fully reconstituted angioarchitecture (i.e.: the sum of signals from all caliber classes). (**c**) is the rendering of the structures excluded from analysis. They are not necessarily contiguous and can be connected with vessels of lower caliber that have been normally analyzed

2. When analyzing different conditions (e.g.: A, untreated tumors; B, tumors treated with substance X), it is not necessary to separate samples into groups given that each sample is elaborated independently from the others. This is true even when recovering signals from original binary stacks, provided no alteration is done, manually. The choice to group samples together according to condition is however an advantage for the human analyst.

3. Sample normalization copying the normalization volume inside each sample folder, assures that partial analyses can be integrated successfully with others analyses carrying additional samples, PROVIDED no new sample shows a greater signal amount (greater Vol%). In this last case, all the samples—the new and the old ones—must be renormalized and reanalyzed.

3.17 Statistical Analyses for Comparison of Angioarchitectures

1. When analyzing microangioarchitectures from any given tissue it is impossible to figure out a representative vascular status with only a single sample. It is thus necessary to create a statistics of the observed microangioarchitectures in order to compare it with statistics from other tissues or from other physiopathological conditions. In this respect, two approaches are feasible and complementary: comparison of the descriptive parameters and comparison of the LR lines.

2. Comparison of descriptive parameters is straightforward as it resolves in column analysis for two or more conditions and can be treated accordingly. As a rule, it is better to perform a one-way ANOVA followed by a specific post hoc test such as Bonferroni's multiple comparisons. Problems can arise in the presence of outliers, given that spatial dispersion values do not follow a Gaussian distribution. In this case, it is better to carry out a nonparametric Kruskal–Wallis test followed by a Dunn post-hoc test.

3. Comparison of LR lines. Using your preferred statistical program (e.g. Prism) create a XY table with two or more conditions and with a number of input replicates equal to the maximal number of samples analyzed among all conditions (e.g. Condition A and condition B, four controls and six treated samples would define a table with 10 columns and 70 rows, see below). Then you sequentially add in the X column blocks of values corresponding to the Vol% values from ordered PRAs of different samples; one block of seven values for each sample. At this point, insert spatial dispersion values (e.g. nHv 95% values) in the appropriate cell defined by row and column for each PRA of each sample. Given the one-to-one correspondence between X and Y values, the table will look essentially empty if not for a diagonal series of blocks of filled cells.

4. Turning to analyses, you will select linear regression and ask for comparison between a couple of conditions testing for differences. The program will use the F test to calculate differences in slope between the two groups and eventually differences in elevations while compensating for void cells and samples. Graphically you can ask to show the 95% confidence intervals for each LR, giving graphical support to differences resulting from statistical tests.

4 Notes

1. Compiling of "Spatial_Dispersion.java" text to an appropriate Plugin is possible using the same ImageJ application. Select "Plugins > Compile and Run . . ." command indicating a saved version of the ".java" text to be compiled. Then save the resulting ".class" file in an appropriate position in order to be present in the Plugins collection shown in the lower half of ImageJ Plugin menu. To this aim select the ImageJ or Fiji folder and the Plugins subfolder. Move your plugin inside this subfolder as such, or create an additional sub-subfolder to store it.

2. In the case your imaging instrument is exporting color images as .czi files, it is important to note that raw ImageJ is unable to process this kind of image format. Therefore you are referred to Fiji to open .czi files and turn them into grayscale images.

3. Beware that file listing can be different between Computer System application and ImageJ. This is especially true when using numerical names, as ImageJ evaluates a name symbol after symbol ignoring the overall value. Thus, 45 appears to occur before 6, falsifying image order and leading to disordered stacks after importing.

4. Usually, motorized microscopes perform z-step acquisition with a fixed increment, independent from image resolution. Report the increment in the depth field of the Property panel according to the stated unit of length.

5. Although output images can be saved according to personal tastes, in some cases macros expect images to be strictly organized in nested subfolders. It is thus advisable to follow a precise rationale for storing resulting images. Authors use a sequence of alphabetically listed folders to group resulting images after each specific procedural step. In this respect the authors recommend the use of self-explanatory names such as a-input images, b-isotropic stacks, c-binary stacks, d-filled masks, e-caliber masks, f-caliber binary stacks, g-partially reconstructed angioarchitectures, h-in analysis, i-results, and j-data elaboration.

6. Saturated voxels. The preset value is a good compromise and is calculated slice after slice unless the stack histogram is used.

7. Equalization. This procedure confers equal representativeness to all histogram areas. In the experience of the authors, this is not necessary to obtain a good contrast and it is often counterproductive.

8. Stack histogram. Upon selection, signals from all the slices concur to the final histogram that is used to define the threshold for saturated voxels. This option is to be preferred unless you observe large inhomogeneity in contrast and brightness along your stack. In this last case, a local, slice by slice histogram can perform better to obtain an optimized contrast.

9. A possibility to combine Default and Renji Entropy thresholding methods relies on z-projection. Rationale for such a possibility assumes that once a sound signal is mapped in a slice, other signals mapping in the same position in different slices have a higher chance to be true signal and not background noise. The procedure sees (a) thresholding a copy of the grayscale stack by the stricter method (Renji Entropy), (b) building a z-projection of the resulting binary stack, (c) thresholding a copy of the grayscale stack by the permissive method, (d) intersecting the permissive binary stack with the z-projection from the restrictive binary stack, and (e) saving the results as a thresholded binary stack.

10. This macro tries to transform hollow vessels in solid pipes in three steps:

    (a) Will try to close holes in the vessel wall dilating and eroding the image by 3D convolution with alternated cross and square kernels according to the number of cycles defined in the Dialog window. If dilating edges fuse, they will not be successively eroded and this operation should lead to the closure of holes with radius smaller than the number of dilating cycles.

    (b) Will execute the ImageJ command "Fill Holes" for every slice in the three Cartesian orientation of the stack. At the end, it will execute a further "Fill Holes" for every slice in standard orientation to ensure to fill new cavities created after the first series of commands.

    (c) Will classify areas of added signals according to a user defined threshold, discarding those larger than the threshold itself. This approach was added in order to avoid filling intervascular spaces instead of intra-vascular funnels. The value of the threshold is variable according to the angioarchitecture and must be defined by the experimenter in order to limit possible artefacts.

This fill-in procedure is ineffective in the presence of largely incomplete staining preventing definition of vessel walls. It is also ineffective in vessels close to borders and bisected by volume boundaries.

11. Our procedure does not allow choosing an arbitrary size as the lower limit in the calibration of vessels because this value is preset to zero. To provide a minimum of flexibility, it is however possible to define the limits of the lowest caliber class choosing among a set of five values. This arrangement is useful because our analysis will work at best when signal voxels are well distributed among all different classes of calibers. Unfortunately, this is not always the case both among small and large vessels. Therefore, the goal is to adapt the boundaries of the different classes of calibers to cover the distribution of the different vascular sizes starting from the smallest available.

    Speaking in pixel values this means that the maximal cross-section that can be analyzed by our macro lies in the 1024–2048 pixel range. That means cross-sections 8 times larger than those analyzed using the conditions selected for this example.

12. In the case of stacks of rectangular images it is not possible to directly use our procedure, but two solutions are possible in order to prepare samples for the analysis:

    (a) crop the volume to the maximal square-based volume (it is not necessary to have a cube, but the volume should show an even number of slices).

    (b) divide the volume in two partially duplicated, square-based volumes obtaining two partially dependent samples from the original input images. This last approach can be useful if you suppose to get different results according to which volume you are analyzing and do not want to lose the information.

    An alternative approach would see a volume expansion to a square-based volume encompassing the whole rectangular volume. The outputs will then be masked to recover the original parallelepiped volume. Given that this approach needs to be implemented in all steps down to the final analysis, it was not developed and will not be further discussed in this occasion.

13. The macro developed to recover and classify original signals according to caliber masks implements two main controls in order to forbid the use of unwanted masks. First of all the macro checks that the number of the original binary volumes to decompose (in this case 1) is equal to the number of the folders grouping the caliber masks to be used.

    Then, the macro performs a congruence control on the names of the files: the title of the original image file to decompose, up to the first full stop (e.g., binadjG12S_Ctrl2.*2-iso.tif*) must be equal to the name of the subfolder grouping the masks with the different calibers, devoid of the very first five characters and up to the first full stop (e.g.: *bp600*binadjG12S_Ctrl2.*2-iso.tif*). The final folder identifying the PRAs will be named after this title, therefore introduction of full stops in sample names is not recommended because of potential overwriting of results and/or macro failure.

14. The rationale for the nested folders is that the folder to be selected according to macro indications groups all the sub-folders grouping the caliber signals divided into 7 equidimensional volumes. In the case of this analysis we have a single sample, hence a single subfolder, that must be placed inside an otherwise void parent folder. As for the 7 caliber volumes they must be organized into an ascending order from smallest to largest. The output of the previous macro is organized in order to cope with these requirements.

15. The major problem trying to organize the sequence of images at different steps in the cascade of preparatory events leading to final analysis is to identify any single file unequivocally. This goal is disturbed by the need to uniform file inputs and outputs to allow for automatic analysis. The authors devised a hierarchical organization in which the meaning of an image is defined by its name and by the path to its location. In this context, short acronyms contribute to identify the level of elaboration to which the image has progressed, whereas numbers often represent thresholds characterizing the class of caliber or the PRA relative to the image. As an additional point, the authors make large use of alphabetical symbols to obtain program-independent organized lists, as numbers are not reliable to the scope. Changes in this nomenclature are irrelevant as far as a similar organization is maintained, but authors warn the user against inconsiderate changes.

16. If everything went smoothly, the volume carrying more signal will necessarily be called TD0-xxx with xxx being the largest threshold used in calibrating vessels (256 in our example). Given that all these volumes coming from different samples share the same name, they cannot be grouped in the same folder without renaming them. However, in this case the analysis of a single folder would provide the name of the normalizing volume.

17. After definition of the volume which can be used to normalize samples, it must be duplicated, renamed to a general name (e.g., z-norm.tif) and copied in the folders of the samples under analysis, except to its own. In order to be able to remove easily the normalizing results from the results table, it is better to rename the normalizing volume in order to give it the first or the last name in the sequence of PRAs to be analyzed.

18. Unfortunately, the progression bar is relative to the analysis of seven PRAs from a single sample plus the normalizer volume and is not very useful in estimating the end of the overall analysis.

19. Values of R^2 for single near-linear lines usually range from 0.95 to 0.90. In some occasions you can observe values down to 0.85, but it will be pretty unusual if you get values lower than 0.80. In presence of such values it is necessary to investigate the shape of the line, as the main cause for such a behavior might reside in the absence of signals from specific classes of calibers. In these conditions, it might be helpful to modify thresholds in order to get a better coverage of vascular calibers.

5 Rationale for Plugin "Spatial Dispersion" [https://github.com/nHv95/Vessel_analysis/Spatial_Dispersion.java]

The plugin operates on a folder containing a subset of folders, each containing a number of stacks with exactly the same X-, Y-, and Z-dimensions.

After collecting information about the path to the input and target folders and after assessing the starting time, the plugin begins the operations on a subfolder basis.

At first it executes a dimensional control on all the image stacks and, if successful, calculates the maximal percent Volume (Vol%) occupied by signal voxels.

Normalization is carried out only on volumes present inside the subfolder under analysis, but given that the normalizing volume has been copied to all subfolders, the whole set of subfolders results normalized and comparable.

At this point, the plugin opens the first stack in the list and creates a duplicate which will play the dilated output. If the input stack presents some signal (i.e., it is not completely blanc) the plugin starts dilating the signals until they will exceed 99.5% of the available volume. Alternatively, the plugin will stop the main cycle in the unlikely case that the number of cycles exceeds the number of 500.

Expansion occurs in 3D on a slice basis and regards the ith image but also the contributes to the ith slice deriving from its closest upper and lower neighbors (namely: i, $i - 1$, $i + 1$). These contributes will change according to the actual dilation scheme (3D cross or cubic) automatically selected as a function of the cycle number. If input neighbor slices do not exist they are created as single void slices with the appropriate dimensions. Contributes are then integrated into the dilated ith slice.

- 3D cross dilation for slice neighbors $i + 1$ and $i - 1$ is equal to a not dilated copy of the input i slice. At a difference, the i image is convolved by a 3×3 cross kernel.

- In the cubic dilation scheme, dilated contributes for all the three slices are equal to an input i, $i - 1$ and $i + 1$ slices, respectively convolved by a 3×3 square kernel.

The plugin calculates the occupied Vol% in the dilated slice and stores the value in an array indexed according to slices, before modifying the output image stack adding the new i slice.

After dilating all the slices, stacks are swapped and the dilated stack becomes the new input stack whereas the old input stack is discarded. At this point, the plugin sums the Vol% occupied by signals in all slices of the dilated volume and stores it in an array indexed according to cycle. This step is necessary because rhombicuboctahedral expansion is performed as a sum of two different expansion (3D cross and cubic). Therefore, the occupied volume will be precise only every 2 cycles, whereas intermediate values will be approximated by linear interpolation between the two precise extremes after completing every second cycle.

With the end of the dilation cycle, the plugin calculates the raw Hv indexes reading the array storing the Vol% values referring to every dilation cycle, starting from the lowest. When the value will exceed 90%, 95% or 99% the plugin will interpolate the relative number of cycles by linear regression defining the raw Hv indexes. The plugin will calculate also the number of cycles needed to reach the normalizing value (N) with a similar approach, starting from the initial Vol%. Again, when the dilated values will exceed N, the plugin will calculate the number of cycles needed to reach normalization by linear regression. The normalized set of nHv indexes is obtained subtracting the normalizing value from the raw Hv indexes.

The results are appended in the output file, one for subfolder, and the plugin will turn to the analysis of a new stack, processing subfolder after subfolder.

Acknowledgments

This work was supported by Fondazione Cariplo grant no. 2016-0570 to AG and also supported by "Regione Lombardia, Progetto GenePark (#149065)".

References

1. Potente M, Mäkinen T (2017) Vascular heterogeneity and specialization in development and disease. Nat Rev Mol Cell Biol 18:477–494. https://doi.org/10.1038/nrm.2017.36

2. Bullitt E, Rahman FN, Smith JK, Kim E, Zeng D, Katz LM et al (2009) The effect of exercise on the cerebral vasculature of healthy aged subjects as visualized by MR angiography. Am J Neuroradiol 30:1857–1863. https://doi.org/10.3174/ajnr.A1695

3. Carmeliet P, Jain RK (2011) Molecular mechanisms and clinical applications of angiogenesis. Nature 473:298–307. https://doi.org/10.1038/nature10144

4. Weidner N, Semple JP, Welch WR, Folkman J (1991) Tumor angiogenesis and metastasis-correlation in invasive breast carcinoma. N Engl J Med 324:1–8. https://doi.org/10.1056/NEJM199101033240101

5. Nico B, Benagiano V, Mangieri D, Maruotti N, Vacca A, Ribatti D (2008) Evaluation of microvascular density in tumors: pro and contra. Histol Histopathol 23:601–607. https://doi.org/10.14670/HH-23.601

6. Lang S, Müller B, Dominietto MD et al (2012) Three-dimensional quantification of capillary networks in healthy and cancerous tissues of two mice. Microvasc Res 84:314–322. https://doi.org/10.1016/j.mvr.2012.07.002

7. Hathout L, Do HM (2012) Vascular tortuosity: a mathematical modeling perspective. J Physiol Sci 62:133–145. https://doi.org/10.1007/s12576-011-0191-6

8. Shelton SE, Lee YZ, Lee M et al (2015) Quantification of microvascular tortuosity during tumor evolution using acoustic angiography. Ultrasound Med Biol 41:1896–1904. https://doi.org/10.1016/j.ultrasmedbio.2015.02.012

9. Downey CM, Singla AK, Villemaire ML, Buie HR, Boyd SK, Jirik FR (2012) Quantitative ex-vivo micro-computed tomographic imaging of blood vessels and necrotic regions within tumors. PLoS One 7:e41685. https://doi.org/10.1371/journal.pone.0041685

10. Zudaire E, Gambardella L, Kurcz C, Vermeren S (2011) A computational tool for quantitative analysis of vascular networks. PLoS One 6:e27385. https://doi.org/10.1371/journal.pone.0027385

11. Tan H, Wang D, Li R et al (2016) A robust method for high-precision quantification of the complex three-dimensional vasculatures acquired by X-ray microtomography. J Synchrotron Radiat 23(Pt 5):1216–1226. https://doi.org/10.1107/S1600577516011498

12. Pabst AM, Ackermann M, Wagner W, Haberthür D, Ziebart T, Konerding MA (2014) Imaging angiogenesis: perspectives and opportunities in tumour research - a method display. J Craniomaxillofac Surg 42:915–923. https://doi.org/10.1016/j.jcms.2014.01.010

13. Nagy JA, Chang SH, Shih SC, Dvorak AM, Dvorak HF (2010) Heterogeneity of the tumor vasculature. Semin Thromb Hemost 36:321–331. https://doi.org/10.1055/s-0030-1253454

14. Gaehtgens P (1991) Heterogeneity of capillary perfusion. Blood Vessels 28:197–200

15. Danielson P (1980) Euclidean distance mapping. Comp Graph Image Process 14:227–248

16. Schneider CA, Rasband WS, Eliceiri KW (2012) NIH Image to ImageJ: 25 years of image analysis. Nat Methods 9:671–675. PMID 22930834

17. Schindelin J, Arganda-Carreras I, Frise E, Kaynig V, Longair M, Pietzsch T et al (2012) Fiji: an open-source platform for biological-image analysis. Nat Methods 9:676–682., PMID 22743772. https://doi.org/10.1038/nmeth.2019

18. Ragnelmann I (1993) The euclidean distance transformation in arbitrary dimensions. Pattern Recogn Lett 14:883–888

19. Jones MW, Bærentzen JA, Sramek M (2006) 3D distance fields: a survey of techniques and applications. IEEE Trans Vis Comput Graph 12:581–599

20. Rodbard S (1975) Vascular caliber. Cardiology 60:4–49. https://doi.org/10.1159/000169701

21. Righi M, Giacomini A, Cleris L, Carlo-Stella C (2013) (3)D [corrected] quantification of tumor vasculature in lymphoma xenografts in NOD/SCID mice allows to detect differences among vascular-targeted therapies. PLoS One 8:e59691. https://doi.org/10.1371/journal.pone.0059691

22. Righi M, Locatelli SL, Carlo-Stella C, Presta M, Giacomini A (2018) Vascular amounts and dispersion of caliber-classified vessels as key parameters to quantitate 3D micro-angioarchitectures in multiple myeloma experimental tumors. Sci Rep 8:17520. https://doi.org/10.1038/s41598-018-35788-4

23. Righi M, Belleri M, Presta M, Giacomini A (2019) Quantification of 3D brain microangioarchitectures in an animal model of

Krabbe Disease. Int J Mol Sci 20:2384. https://doi.org/10.3390/ijms20102384

24. Meijering EHW, Niessen WJ, Viergever MA (2001) Quantitative evaluation of convolution-based methods for medical image interpolation. Med Image Anal 5:111–126

25. https://imagej.nih.gov/ij/developer/macro/macros.html

26. https://imagej.nih.gov/ij/docs/pdfs/tutorial10.pdf

27. Rybak JN, Ettorre A, Kaissling B, Giavazzi R, Neri D, Elia G (2005) In vivo protein biotinylation for identification of organ-specific antigens accessible from the vasculature. Nat Methods 2:291–298

28. Lavazza C, Carlo-Stella C, Giacomini A, Cleris L, Righi M, Sia D et al (2010) Human CD34+ cells engineered to express membrane-bound tumor necrosis factor-related apoptosis-inducing ligand target both tumor cells and tumor vasculature. Blood 115:2231–2240

Chapter 13

A Xenograft Model for Venous Malformation

Jillian Goines and Elisa Boscolo

Abstract

Xenograft models allow for an in vivo approach to monitor cellular functions within the context of a host microenvironment. Here we describe a protocol to generate a xenograft model of venous malformation (VM) based on the use of human umbilical vein endothelial cells (HUVEC) expressing a constitutive active form of the endothelial tyrosine kinase receptor TEK (TIE2 p.L914F) or patient-derived EC containing TIE2 and/or PIK3CA gene mutations. Hyperactive somatic *TIE2* and *PIK3CA* mutations are a common hallmark of VM in patient lesions. The EC are injected subcutaneously on the back of athymic nude mice to generate ectatic vascular channels and recapitulate histopathological features of VM patient tissue histology. Lesion plugs with TIE2/PIK3CA-mutant EC are visibly vascularized within 7–9 days of subcutaneous injection, making this a great tool to study venous malformation.

Key words Venous malformation, Xenograft model, Endothelial cells, TIE2

1 Introduction

The establishment of the vasculature includes two main processes termed vasculogenesis and angiogenesis. Vasculogenesis is the "de novo" formation of blood vessels from precursor cells, while angiogenesis entails the subsequent sprouting from the preexisting vessels [1]. Defects during the angiogenic process can lead to vascular anomalies such as venous malformation (VM) which is characterized by the development of greatly dilated, ectatic veins [2–4].

Normal and pathological angiogenesis encompasses several critical cellular functions such as endothelial cell adhesion, extracellular matrix (ECM) remodeling, proliferation, migration, and lumen formation [5]. While in vitro 2D cultures of endothelial cells (EC) provide a platform to investigate each of these cellular properties, the combined interplay of EC functions can be best assessed in an in vivo setting. In vivo models of vascular anomalies are based on 1-transgenic animals expressing the genetic defect identified in patients and on 2-xenografts consisting of transplantation or injection of human patient tissue or cells. In the vascular

Domenico Ribatti (ed.), *Vascular Morphogenesis: Methods and Protocols*, Methods in Molecular Biology, vol. 2206, https://doi.org/10.1007/978-1-0716-0916-3_13, © Springer Science+Business Media, LLC, part of Springer Nature 2021

anomaly field, transgenic mouse models are primarily based on dominant negative [6] or constitutively active genetic mutations [7–11]. Transgenic mouse models of VM are based on the ubiquitous (CAG-CreERT2), epithelial/endothelial (Sprr2f+), or embryonic mesoderm–restricted expression of the human PIK3CA p. H1047R [7, 8]. While a transgenic murine model of VM based on *TIE2* mutations has not been reported yet, expression of TIE2 p.R849W in zebrafish resulted in formation of vascular lesions resembling human VM lesions [12]. Transgenic models are great tools for investigating the whole body or tissue-specific effects of different genetic mutations or gene loss of function; however, this can result in widespread pathological vasculature and early lethality. Current transgenic models fail to take into account the sporadic nature of the mutational events found in patient tissue, and may not exactly recapitulate the effects that result in localized vascular lesions. Conversely, xenograft models allow for the use of patient-derived tissue or cells to generate patient-specific platforms some of which are widely used in the cancer field to test drug treatments for precision medicine.

Xenografts are based on the use of immunocompromised mice that do not reject human cells. Commonly used mouse strains are: athymic nude (nu/nu) mice, severely compromised immunodeficient (SCID; Prkdc^{scid}) mice, and recombination-activating gene 2 (Rag2)-knockout mice. To develop a mouse model with a human immune system, humanized mice can be generated by the administration in NSG (NOD scid gamma *mouse*) mice of human immune cells before implantation of the xenograft [13].

Patient-derived xenografts (PDX) based on tissue or organoid transplantation are widely used to study tumor propagation and to determine drug dosing, treatment schedules, and routes of administration that maximize antitumor efficacy and therapeutic window [14]. Subcutaneous inoculation of different cell types is also used to generate xenografts for the study of tumor growth, metastasis, and vascular anomalies or for regenerative medicine. The advantages of this method are the direct access for implantation and ease of monitoring the tumor/lesion growth. Furthermore, a recent study showed that injection of unassembled vascular cells (i.e., cell suspension) allows for anastomosis with the host vasculature and subsequent perfusion while preassembled vascular networks do not engraft efficiently [15]. In most cases, single cell suspension of normal EC such as human umbilical vein endothelial cells (HUVEC) or cord blood endothelial colony-forming cells (cbECFC) requires a support cell type such as smooth muscle cells or bone marrow mesenchymal progenitor cells (bmMPC) to assemble into blood vessels and anastomose with the host vascular network [16, 17]. Here we describe a protocol in which EC isolated from lesions in VM patients and expressing a mutated constitutive

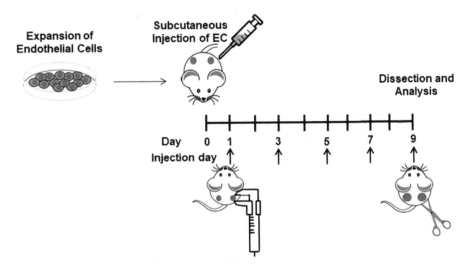

Fig. 1 Schematic of cell injection in mouse and timeline until dissection. Endothelial cells are expanded to desired number of cells prior to injection day. Subcutaneous injection of EC on day 0 is followed by initial lesion measurement, day 1. Lesions are measured every other day, as signified by arrows, using a caliper through experimental day 9. Lesions are dissected and processed for tissue analysis on day 9

active form of *TIE2* or *PIK3CA* can undergo significant morphogenesis and generate perfused pathological ectatic vessels even when injected alone [18–20].

To promote EC morphogenesis in the xenograft, the injected vascular cells are mixed with an ECM component solution such as Fibronectin and Collagen, or a mixture of ECM and growth factors such as Matrigel® (matrix preparation enriched in collagen IV, nidogen and laminin that is derived from the mouse Engelbreth-Holm–Swarm sarcoma) or the peptide hydrogel PuraMatrix® [21, 22].

In this protocol, we present methods for establishing a patient-derived xenograft in mice based on injection of EC isolated from VM lesions in patients or HUVEC expressing TIE2 p.L914F, the most frequent mutation found in individuals affected by VM (Fig. 1).

2   Materials

All buffers and solutions are prepared at room temperature (RT) and stored at 4 °C (unless indicated otherwise). Consumable equipment, such as pipette tips, needles, syringes, and microtubes should be sterilized prior to use. All reagents are prepared in a clean biosafety cabinet and waste disposed of according to biohazardous waste disposal regulations.

2.1 Cell Culture

1. Endothelial cells (EC): HUVEC-TIE2-L914F and patient-derived EC (CD31+) (*see* **Note 1**).

2. Endothelial growth medium (10% FBS/EGM-2): Endothelial Cell Growth Medium-2 (EGM™-2), Fetal Bovine Serum (FBS): Remove 50 mL of endothelial basal medium (EBM-2) from bottle and add 50 mL of FBS. Thaw (*see* **Note 2**) and add SingleQuots™ growth factor supplement to bottle, except for hydrocortisone. Add 5 mL of 100× penicillin–streptomycin–L-glutamine. Gently mix and sterile-filter through a 0.2 μm pore-sized bottle filter into an autoclaved Wheaton bottle. Store at 4 °C (or −20 °C) in 50 mL conical tube aliquots (*see* **Note 3**).

3. Sodium carbonate coating buffer (Na_2CO_3 0.1 M pH 9.4): Weigh 5.3 g of sodium carbonate in 500 mL of deionized water. Adjust pH to 9.4 and filter through a 0.2 μm bottle top filter into an autoclaved Wheaton bottle. Store at RT.

4. Human plasma fibronectin purified protein.

5. PBS: Sterile 1× phosphate buffered solution.

6. 15/50 mL conical tubes.

7. 100 × 20 mm tissue culture–treated plates.

8. 145 × 20 mm tissue culture–treated plates.

9. Trypsin solution: 0.05% trypsin–0.5 μM EDTA.

10. Trypan Blue Solution, 0.4%.

11. Hausser Scientific Bright-Line™ Counting Chambers.

2.2 Subcutaneous Injection of Cells into Mouse

1. Mice, 5–6 week old Hsd:Athymic Nude Foxn1^{nu}.

2. Needles, 26 G × 5/8 in. Sub-Q sterile needles.

3. Syringes, 1 mL with Luer Lock.

4. Matrigel: Corning™ Matrigel® Basement Membrane Matrix (Phenol Red-Free).

5. Ice bucket filled with ice.

6. Aluminum foil.

7. Isoflurane.

8. Anesthesia machine with nose cone attachment.

9. Ear punch.

2.3 Lesion Plug Measurements and Collection

1. Caliper.

2. Scale to weigh mice.

3. Dissection tools, including scissors and forceps.

4. Cutting board.

5. Seventy percent ethanol (EtOH) in deionized water.

6. Camera stage with camera.

7. Ruler.

8. 10% formalin buffered solution.

2.4 Tissue Staining

1. Positive-charged glass slides.

2. Staining slide holder and containers.

3. Xylene.

4. Ethanol (100%, 90%, 80%, 70%) in deionized water.

5. Hematoxylin (Gill's Formula).

6. Eosin Y (alcohol-based).

7. Antigen retrieval solution: Tris–EDTA, pH 9.0. Weigh 0.6 g of Trizma base and 1 mL of 0.5 M EDTA to 500 mL of deionized water. Adjust pH to 9.0. Add 250 μL of Tween 20. Store at room temperature (RT).

8. Humidifying slide chamber.

9. Liquid-repellent Super PAP pen.

10. Blocking buffer: 5% normal horse serum in PBS.

11. Biotinylated *Ulex europaeus* agglutinin-1 (UEA-1).

12. 3% hydrogen peroxide solution: To prepare, add 1 mL of 30% hydrogen peroxide to 9 mL of deionized water.

13. Streptavidin horseradish peroxidase (HRP)-conjugated Antibody.

14. DAB: 3,3′Diaminobenzidine Reagent. Peroxidase Substrate Kit. Prepare according to the manufacturer's recommendations.

15. Permount Mounting Medium.

16. Coverslips.

17. Brightfield microscope with 20× objective for imaging.

18. ImageJ Software.

3 Methods

Carry out procedures in a biosafety cabinet, unless otherwise specified. All procedures must be approved under IBC Biosafety Level 2 regulations and protocol. All mouse procedures must be performed under an approved IACUC protocol and regulations observed. This protocol should be completed as quickly and efficiently as possible to minimize cell exposure to stress.

3.1 Cell Plating and Expansion

It is important to culture and expand more EC than what is needed for the injection. It is recommended by our laboratory to count cells during passaging step to ensure cells are growing and to properly plan out when to schedule your injection. Here we will describe how to thaw, passage, and maintain EC culture to prepare for in vivo injection of the cells.

1. Plates will need to be ECM-coated to ensure proper cell adhesion. Pipet 5 mL of sodium carbonate coating buffer per 100×20 mm tissue culture–treated plate. Add $1 \ \mu g/cm^2$ of human plasma fibronectin purified protein and gently distribute the liquid onto plate.

2. Incubate plate at $37 \ ^{\circ}C$, 5% CO_2 for 20 min.

3. Aspirate and wash plate with PBS prior to culturing cells.

4. Quickly and completely thaw a cryovial of cells in a $37 \ ^{\circ}C$ water bath. Transfer into a new 15 mL conical tube with 1 mL of endothelial growth medium and pellet cells by centrifugation for 5 min at 400 relative centrifugal force (rcf), RT.

5. Resuspend cell pellet into 10 mL of endothelial growth medium and seed cells into a 100×20 mm tissue culture–treated plate. Change medium every 48 h while cells are recovering.

6. Once cells reach 80–90% confluency, passage cells by seeding 1–1.5×10^4 cells/cm^2 into 145×20 mm tissue culture–treated and fibronectin-coated plates. Add up to a total of 25 mL of endothelial growth medium per plate (*see* **Note 4**). Continue to passage cells until desired cell number has been met. Do not passage cells the day before injection.

3.2 Day Before Injection

1. Prechill syringes, needles, and pipette tips in $-20 \ ^{\circ}C$ freezer overnight.

2. Slowly thaw Matrigel® as per manufacturer instructions. Briefly, thaw a 10 mL bottle immersed on ice and place ice bucket at $4 \ ^{\circ}C$ overnight. Aliquot excess Matrigel® into 1.5 mL microtubes to avoid freeze–thaw cycles.

3.3 Endothelial Cell Preparation and Counting

1. Trypsinize cells using 5 mL of prewarmed trypsin solution per 145×20 mm tissue culture–treated plate. Incubate at $37 \ ^{\circ}C$, 5% CO_2, for 2 min to allow detaching of all cells from plate surface.

2. Neutralize trypsin with 5 mL of endothelial growth medium for every 5 mL of trypsin, transfer to 15 mL conical tube, and pellet cells by centrifugation for 5 min at 400 rcf, RT. Resuspend cells in 10 mL of medium.

3. Mix 20 μL of cells with 20 μL of trypan blue for cell counting using Hausser Counting Chamber [23].

4. Determine the volume needed to obtain 2.5×10^6 cells per injection (*see* **Note 5**).

5. Transfer this volume into a new 15/50 mL conical tube and pellet cells by centrifugation for 5 min at 400 rcf, RT.

6. Aspirate the supernatant, not disturbing the cell pellet (*see* **Note 6**). Chill on ice until ready to resuspend in Matrigel®.

3.4 Syringe Preparation

1. Prepare syringes with needles over ice to prevent solidification of Matrigel®.

2. Dislodge cell pellet by gently tapping the 15 mL conical tube. Resuspend the cell pellet with 200 μL of Matrigel® per injection on ice (*see* **Note 7**).

3. Using 1 mL pipette and 1 mL syringes, simultaneously pipet Matrigel®–cell mixture into syringe opening by suction force while pulling plunger of syringe. Luer lock a 26 G × 5/8 in. sterile needle to the syringe and keep prepared syringes flat on ice prior to injection.

3.5 Subcutaneous Injection into Mouse

1. Prepare anesthesia machine of choice, according to the manufacturer's protocol, and attach nose cone (*see* **Note 8**). Ensure mice are properly anesthetized before subcutaneous injection.

2. Gently roll prepared syringe with needle to reincorporate any settled cells and flick bubbles to the needle end of syringe. Uncap needle and push plunger to distribute mixture through the needle.

3. Pinch and create "tent-like" structure on right back side of mouse hind leg and insert needle subcutaneously right under the skin (*see* **Note 9**). Holding needle at 45° angle carefully inject 200 μL of the cell–Matrigel® mixture to create a small spherical mass (Fig. 2). This procedure can be repeated on the left back side of mouse hind leg. Discard needle and syringe into an appropriate Category 1 sharps container.

4. Record mouse weight using a scale (*see* **Note 10**), ear tag the mouse, and return to cage. Monitor mouse to ensure it returns to normal activity.

3.6 Lesion Growth Monitoring

1. Using a caliper, record lesion area by measuring the length of the long and short sides of the plug as shown in Fig. 3. Note lesion size, color, and any abnormal affects such as bleeding, bruising, or discoloration from previous color record.

2. Measurements are read every other day up until collection day (*see* **Note 11**).

3.7 Tissue Collection and Processing

1. Prior to lesion tissue collection, euthanize mice by CO_2 per IACUC protocol.

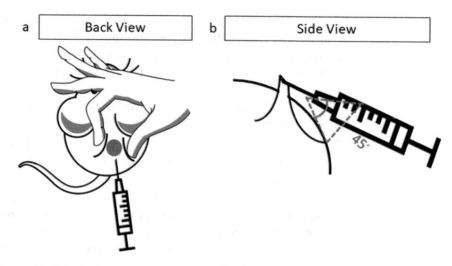

Fig. 2 Schematic of in vivo injection procedure. Skin on the backside of the mouse is pinched using forefinger and thumb to create a tent-like structure. Next, carefully insert needle into subcutaneous layer and release skin of mouse (**a**). From the side, there should be a 45° angle created between the mouse backside and needle when injecting mixture subcutaneously into skin (**b**)

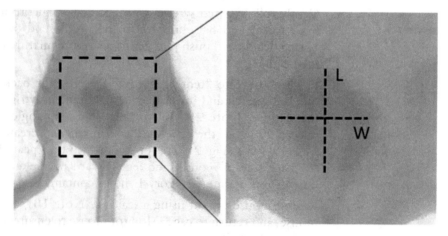

Fig. 3 Caliper measurement of lesion on mouse. Lesion will form a circular structure on the back. Measure lesion size with a caliper by recording the length (L) and the width (W)

2. Spray lesion tissue area with 70% EtOH and carefully dissect plug from hind legs with the use of scissors. Leave skin on plug to prevent altering the lesion mass.

3. Set up a camera stage with a camera. Align plugs onto a cutting board aside a ruler. Take an image of plugs to record gross vascularity of lesions.

4. Fix tissue by placing the lesion plug in 10% formalin overnight at RT. The next day, replace fixative with PBS.

5. Process tissue for paraffin embedding (*see* **Note 12**).

3.8 Staining of Lesion Sections

We stain the lesions for Hematoxylin and Eosin (H&E, *see* Subheading 3.8.1) to detect vascular structures of lesions. We further stain lesions with a human endothelial-specific lectin marker *Ulex europaeus* agglutinin-1 (UEA-1, *see* Subheading 3.8.2) to confirm that vascular structures are formed by human-derived EC.

1. Prepare tissue slides by cutting 5 μm paraffin tissue sections (*see* **Note 13**).

2. Melt wax at 60 °C for 1 h prior to staining.

3. Fumes from xylene and EtOH are toxic. Ensure that entire staining procedure is performed in a well-ventilated area and chemical fume flow hood to minimize fume exposure. De-paraffinize and rehydrate tissue by sequentially incubating slide in xylene for 10 min, 100% EtOH for 5 min, 90% EtOH for 3 min, and 80% EtOH for 3 min. Rinse slide in deionized water for 5 min.

3.8.1 Hematoxylin and Eosin (H&E)

1. Incubate sections in hematoxylin for 2 min. Using a beaker and slide holder, rinse in a sink by a steady stream of tap water until water is clear.

2. Dehydrate slides by incubating tissue sequentially in 70% EtOH for 1 min, 80% EtOH for 1 min, 90% EtOH for 1 min, 100% EtOH for 1 min, and another fresh 100% EtOH for 1 min.

3. Stain sections in Eosin Y for 30 s. Immediately following, rinse in 100% EtOH. Rinse in fresh 100% EtOH until solution is clear.

4. Incubate slide in xylene for 2 min. Let slides dry for 5–10 min under the fume hood.

5. Pipet a drop of Permount Mounting medium per lesion section. At 45° angle, lower coverslip to evenly distribute mounting medium and avoid bubble production under coverslip. Allow slides to dry overnight before imaging.

3.8.2 Ulex europaeus Agglutinin-1 (UEA-1) Immunohistochemistry (IHC)

1. Incubate tissue slide in a beaker with antigen retrieval solution, stirring on a heating block, for 20 min at 95 °C.

2. Remove beaker from heating block, allow solution to cool to 35 °C, then wash in PBS for 3 min.

3. Using a liquid-repellent Super PAP pen, draw a small circle around tissue section to be stained.

4. Incubate slide sections in blocking buffer for 30 min, RT.

5. Prepare a UEA-1 working solution by diluting 20 μg/mL of Biotinylated UEA-1 in blocking buffer. Pipet 50–100 μL of UEA-1 working solution per section and incubate for 1 h at RT in a humidifying chamber (*see* **Note 14**).

6. Wash slides two times in PBS for 3 min each.

7. Peroxidase quench slide sections in 3% hydrogen peroxide for 5 min, RT.

8. Wash slides two times in PBS for 3 min each.

9. Prepare 5 μg/mL of Streptavidin HRP-conjugated in blocking buffer. Pipet 50–100 μL per section and incubate for 1 h at RT in a humidifying chamber.

10. Wash slides two times in PBS for 3 min each.

11. Prepare DAB solution and add 50–100 μL per section. Incubate sections for 10–15 min, checking and monitoring for development of stain every 2–5 min. Following staining, wash in PBS three times for 3 min each.

12. Incubate sections in hematoxylin for 3 min. Using a beaker and slide holder, rinse in a sink by a steady stream of tap water until water is clear.

13. Sequentially incubate slide in 80% EtOH for 1 min, 90% EtOH for 1 min, 100% EtOH for 1 min, and xylene for 2 min. Let slides dry for 5–10 min under the fume hood.

14. Pipet a drop of Permount Mounting medium per section at 45° angle, lower coverslip to evenly distribute mounting medium and avoid bubble production under coverslip. Allow slides to dry overnight before imaging.

3.9 Vascular Channel Analysis

A comprehensive and thorough analysis is performed to better understand the vascular channel formation of these xenograft lesions or eventual response to drug treatments. Vascular channels of lesions are quantified by measuring vascular area and vascular density.

1. Four to five images are taken per lesion section with a bright-field microscope at 20× magnification. Each photo is taken in an x-plane pattern within the lesion section to avoid overlap. Take at least one image with a scale bar for pixel-to-millimeter ratio calibration.

2. Images are loaded and stacked on ImageJ per lesion section. Using ImageJ, calibrate the pixels of your scale bar to convert measurements to millimeters by using "Set scale ..." under Analyze tab. This allows software to quantify relative vascular area for one single lesion.

3. A vascular channel is defined as any area that is lined with EC that may contain blood or fluid inside (Fig. 4). Using Freehand selections tool, manually outline vascular channels per image and add to ROI Manager on ImageJ. Do this for all channels on every image in stack.

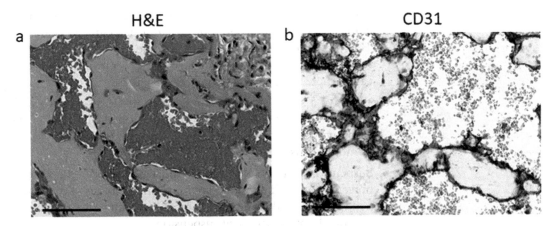

Fig. 4 Staining of lesion plug with HUVEC-TIE2-L914F. Representative images of lesion plug staining of xenograft model. Hematoxylin and eosin (H&E), left (**a**) and immunohistochemistry of endothelial cell marker, anti-human CD31, right (**b**). Scale bar: 100 μm

4. Once all channels are outlined and entered into ROI manage, click on "Measure." Using "Area" measurement, average the total vascular channel area of the lesion over the total area of a single image/field. This will give you the average vascular area per total area for each lesion. Vascular density is measured by manually counting the number of vascular channels per section over the image/field area.

4 Notes

1. Previous publications from our lab have demonstrated that our in vivo model can recapitulate perfused, ectatic VM channel formation using both transfected HUVEC with a TIE2-L914F mutation as well as patient-derived EC [18–20]. Please refer to [19] on how to isolate and generate endothelial cell populations from patient tissue and identify *TIE2* or *PIK3CA* hot-spot mutations.

2. SingleQuots™ must be thawed slowly. Thawing can occur after 30 min at room temperature or up to 2 h at 4 °C. Thawing at higher temperatures is not recommended as it will affect the stability of growth factors.

3. EGM™-2 Endothelial Cell Growth Medium-2: 50 mL aliquots can be stored at 4 °C up to 2 weeks or −20 °C for up to 1 year immediately following preparation.

4. Expansion of cells is based on cell number for optimal seeding and cell proliferation. EC that are growing well may require passaging every 48–72 h. Continue to passage cells on 145 × 20 mm plates until ready to be injected.

5. Any cells left over that will not be used for injection can be cultured at a density of $1–2 \times 10^4$ cells/cm^2 on a coated 100×20 mm or 145×20 mm tissue culture–treated plate to reduce the amount of time cells are left out of incubator in cases where storage of cells is necessary for use later.

6. We leave a small quantity of medium (about 50–70 μL) with pellet to break up cells prior to resuspension in Matrigel®. This helps with the viscosity of the Matrigel® to create a more homogeneous solution of cells for injections.

7. Suspension of cells in Matrigel® must be performed on ice to prevent from solidifying. Avoid creating bubbles while resuspending. Pipet thoroughly to prevent clumps of cells and to obtain a homogenous cell suspension.

8. Once mouse is anesthetized, use nose cone attachment to ensure mouse is properly anesthetized throughout the injection procedure. Anesthesia machine method is preferred to use as little anesthesia as possible and monitor mouse activity quickly following subcutaneous injection. In our laboratory, we use the EZ-Anesthesia Classic System with oxygen flowmeter rate between 1 and 3 and isoflurane vaporizer concentration level between 2 and 3 for our mice.

9. Ensure that needle is only skin deep by releasing pinched skin. Avoid poking the hind leg muscle to prevent injection of plug into muscle. It is our goal to study vessel development that occurs within the Matrigel® plug with minimal assistance infiltration from mouse endothelium.

10. As per our IACUC protocol, we record mouse weight to ensure that our lesion size does not become a burden to the mouse's health as the lesion grows over time.

11. Growing rate of lesions may depend on the type and number of EC injected. Xenograft lesion size at Day 1 is about 80–100 mm^2. In the case for venous malformation EC, 9 days is a sufficient amount of time to visualize enlarged vascular channels of human origin in the center of the plug, with minimal assistance from mouse endothelium, typically measuring up to 150–200 mm^2.

12. We use the Pathology Core at Cincinnati Children's Hospital Medical Center to process our tissue for paraffin blocks.

13. To analyze vasculature inside of lesion plug, we cut 5 μm-thick sections sequentially. It is recommended that initial hematoxylin and eosin staining analysis begins 100 μm into the tissue (about 1/3 of the xenograft plug). Moving toward the middle of the lesion is optimal to ensure vessels are formed by the human-derived endothelium rather than the mouse endothelium.

14. We encourage staining of UEA-1 or a human endothelial-specific antibody, such as anti-platelet endothelial cell adhesion molecule (PECAM1 or CD31), to ensure proper xenograft analysis of channels are made by the injected human EC versus mouse EC.

Acknowledgments

Research reported in this chapter was supported by the National Heart, Lung, and Blood Institute, under Award Number R01 HL117952 (E.B.), part of the National Institutes of Health. The content is solely the responsibility of the authors and does not necessarily represent the official views of the National Institutes of Health.

References

1. Folkman J, D'Amore PA (1996) Blood vessel formation: what is its molecular basis? Cell 87 (7):1153–1155

2. Brouillard P, Vikkula M (2007) Genetic causes of vascular malformations. Hum Mol Genet 16 Spec No. 2:R140–R149. https://doi.org/10.1093/hmg/ddm211

3. Dompmartin A, Vikkula M, Boon LM (2010) Venous malformation: update on aetiopathogenesis, diagnosis and management. Phlebology 25(5):224–235. https://doi.org/10.1258/phleb.2009.009041

4. Vikkula M, Boon LM, Carraway KL 3rd, Calvert JT, Diamonti AJ, Goumnerov B, Pasyk KA, Marchuk DA, Warman ML, Cantley LC, Mulliken JB, Olsen BR (1996) Vascular dysmorphogenesis caused by an activating mutation in the receptor tyrosine kinase TIE2. Cell 87(7):1181–1190. S0092-8674(00)81814-0 [pii]

5. Stratman AN, Saunders WB, Sacharidou A, Koh W, Fisher KE, Zawieja DC, Davis MJ, Davis GE (2009) Endothelial cell lumen and vascular guidance tunnel formation requires MT1-MMP-dependent proteolysis in 3-dimensional collagen matrices. Blood 114 (2):237–247. https://doi.org/10.1182/blood-2008-12-196451

6. Crist AM, Lee AR, Patel NR, Westhoff DE, Meadows SM (2018) Vascular deficiency of Smad4 causes arteriovenous malformations: a mouse model of Hereditary Hemorrhagic Telangiectasia. Angiogenesis 21(2):363–380. https://doi.org/10.1007/s10456-018-9602-0

7. Castel P, Carmona FJ, Grego-Bessa J, Berger MF, Viale A, Anderson KV, Bague S, Scaltriti M, Antonescu CR, Baselga E, Baselga J (2016) Somatic PIK3CA mutations as a driver of sporadic venous malformations. Sci Transl Med 8(332):332ra342. https://doi.org/10.1126/scitranslmed.aaf1164

8. Castillo SD, Tzouanacou E, Zaw-Thin M, Berenjeno IM, Parker VE, Chivite I, Mila-Guasch M, Pearce W, Solomon I, Angulo-Urarte A, Figueiredo AM, Dewhurst RE, Knox RG, Clark GR, Scudamore CL, Badar A, Kalber TL, Foster J, Stuckey DJ, David AL, Phillips WA, Lythgoe MF, Wilson V, Semple RK, Sebire NJ, Kinsler VA, Graupera M, Vanhaesebroeck B (2016) Somatic activating mutations in Pik3ca cause sporadic venous malformations in mice and humans. Sci Transl Med 8(332):332ra343. https://doi.org/10.1126/scitranslmed.aad9982

9. di Blasio L, Puliafito A, Gagliardi PA, Comunanza V, Somale D, Chiaverina G, Bussolino F, Primo L (2018) PI3K/mTOR inhibition promotes the regression of experimental vascular malformations driven by PIK3CA-activating mutations. Cell Death Dis 9(2):45. https://doi.org/10.1038/s41419-017-0064-x

10. Rodriguez-Laguna L, Agra N, Ibanez K, Oliva-Molina G, Gordo G, Khurana N, Hominick D, Beato M, Colmenero I, Herranz G, Torres Canizalez JM, Rodriguez Pena R, Vallespin E, Martin-Arenas R, Del Pozo A, Villaverde C,

Bustamante A, Ayuso C, Lapunzina P, Lopez-Gutierrez JC, Dellinger MT, Martinez-Glez V (2019) Somatic activating mutations in PIK3CA cause generalized lymphatic anomaly. J Exp Med 216(2):407–418. https://doi.org/10.1084/jem.20181353

11. Venot Q, Blanc T, Rabia SH, Berteloot L, Ladraa S, Duong JP, Blanc E, Johnson SC, Hoguin C, Boccara O, Sarnacki S, Boddaert N, Pannier S, Martinez F, Magassa S, Yamaguchi J, Knebelmann B, Merville P, Grenier N, Joly D, Cormier-Daire V, Michot C, Bole-Feysot C, Picard A, Soupre V, Lyonnet S, Sadoine J, Slimani L, Chaussain C, Laroche-Raynaud C, Guibaud L, Broissand C, Amiel J, Legendre C, Terzi F, Canaud G (2018) Targeted therapy in patients with PIK3CA-related overgrowth syndrome. Nature 558 (7711):540–546. https://doi.org/10.1038/s41586-018-0217-9

12. Du Z, Ma HL, Zhang ZY, Zheng JW, Wang YA (2018) Transgenic expression of A venous malformation related mutation, TIE2-R849W, significantly induces multiple malformations of zebrafish. Int J Med Sci 15(4):385–394. https://doi.org/10.7150/ijms.23054

13. Okada S, Vaeteewoottacharn K, Kariya R (2019) Application of highly immunocompromised mice for the establishment of patient-derived xenograft (PDX) models. Cells 8(8). https://doi.org/10.3390/cells8080889

14. Byrne AT, Alferez DG, Amant F, Annibali D, Arribas J, Biankin AV, Bruna A, Budinska E, Caldas C, Chang DK, Clarke RB, Clevers H, Coukos G, Dangles-Marie V, Eckhardt SG, Gonzalez-Suarez E, Hermans E, Hidalgo M, Jarzabek MA, de Jong S, Jonkers J, Kemper K, Lanfrancone L, Maelandsmo GM, Marangoni E, Marine JC, Medico E, Norum JH, Palmer HG, Peeper DS, Pelicci PG, Piris-Gimenez A, Roman-Roman S, Rueda OM, Seoane J, Serra V, Soucek L, Vanhecke D, Villanueva A, Vinolo E, Bertotti A, Trusolino L (2017) Interrogating open issues in cancer precision medicine with patient-derived xenografts. Nat Rev Cancer 17(4):254–268. https://doi.org/10.1038/nrc.2016.140

15. Lin RZ, Lee CN, Moreno-Luna R, Neumeyer J, Piekarski B, Zhou P, Moses MA, Sachdev M, Pu WT, Emani S, Melero-Martin JM (2017) Host non-inflammatory neutrophils mediate the engraftment of bioengineered vascular networks. Nat Biomed Eng:1. https://doi.org/10.1038/s41551-017-0081

16. Melero-Martin JM, De Obaldia ME, Kang SY, Khan ZA, Yuan L, Oettgen P, Bischoff J (2008) Engineering robust and functional vascular networks in vivo with human adult and cord blood-derived progenitor cells. Circ Res 103 (2):194–202. https://doi.org/10.1161/CIRCRESAHA.108.178590

17. Melero-Martin JM, Khan ZA, Picard A, Wu X, Paruchuri S, Bischoff J (2007) In vivo vasculogenic potential of human blood-derived endothelial progenitor cells. Blood 109 (11):4761–4768

18. Boscolo E, Limaye N, Huang L, Kang KT, Soblet J, Uebelhoer M, Mendola A, Natynki M, Seront E, Dupont S, Hammer J, Legrand C, Brugnara C, Eklund L, Vikkula M, Bischoff J, Boon LM (2015) Rapamycin improves TIE2-mutated venous malformation in murine model and human subjects. J Clin Invest 125(9):3491–3504. https://doi.org/10.1172/JCI76004

19. Goines J, Li X, Cai Y, Mobberley-Schuman P, Metcalf M, Fishman SJ, Adams DM, Hammill AM, Boscolo E (2018) A xenograft model for venous malformation. Angiogenesis 21 (4):725–735. https://doi.org/10.1007/s10456-018-9624-7

20. Li X, Cai Y, Goines J, Pastura P, Brichta L, Lane A, Le Cras TD, Boscolo E (2019) Ponatinib combined with rapamycin causes regression of murine venous malformation. Arterioscler Thromb Vasc Biol 39 (3):496–512. https://doi.org/10.1161/ATVBAHA.118.312315

21. Allen P, Kang KT, Bischoff J (2015) Rapid onset of perfused blood vessels after implantation of ECFCs and MPCs in collagen, PuraMatrix and fibrin provisional matrices. J Tissue Eng Regen Med 9(5):632–636. https://doi.org/10.1002/term.1803

22. Allen P, Melero-Martin J, Bischoff J (2011) Type I collagen, fibrin and PuraMatrix matrices provide permissive environments for human endothelial and mesenchymal progenitor cells to form neovascular networks. J Tissue Eng Regen Med 5(4):e74–e86. https://doi.org/10.1002/term.389

23. Green MR, Sambrook J (2019) Estimation of cell number by hemocytometry counting. Cold Spring Harb Protoc 2019(11):pdb prot097980. https://doi.org/10.1101/pdb.prot097980

Chapter 14

In Vivo Vascular Network Forming Assay

Hwan D. Kim, Ruei-Zeng Lin, and Juan M. Melero-Martin

Abstract

The capability of forming functional blood vessel networks is critical for the characterization of endothelial cells. In this chapter, we will review a modified in vivo vascular network forming assay by replacing traditional mouse tumor-derived Matrigel with a well-defined collagen–fibrin hydrogel. The assay is reliable and does not require special equipment, surgical procedure, or a skilled person to perform. Moreover, investigators can modify this method on-demand for testing different cell sources, perturbation of gene functions, growth factors, and pharmaceutical molecules, and for the development and investigation of strategies to enhance neovascularization of engineered human tissues and organs.

Key words Vascularization, Endothelial, Vessel formation, Tissue engineering

1 Introduction

The Matrigel plug angiogenesis assay was initially developed as a simple in vivo model to study the growth of new vessels [1]. This assay is widely used as a valuable technique for in vivo screening of pro- and antiangiogenic compounds [2]. Nonetheless, in this procedure, the Matrigel plug contains no added cell, and the assay solely reflects vascularization as the angiogenic ingrowth of surrounding host vessels into the implanted plug. Around 2007, we and others established an alternative in vivo vascular network forming assay by adding vascular cells into the Matrigel plug [3–7]. Cells were harvested from cell culture as single-cell suspensions and mixed in Matrigel for injection, mainly into the subcutaneous space of immunodeficient mice. In this assay, vascular cells rearrange and assemble into new blood vessels after implantation, a process sometimes known as vasculogenesis (i.e., the de novo formation of a primitive vascular network) [8]. Over the last decade, this assay has allowed testing the capability of various cell sources to form blood vessels in vivo [3, 5–7, 9, 10], and studies have consistently established that two major cell types are essential for robust vascularization: endothelial cells (ECs) and perivascular cells [3, 5].

Domenico Ribatti (ed.), *Vascular Morphogenesis: Methods and Protocols*, Methods in Molecular Biology, vol. 2206,
https://doi.org/10.1007/978-1-0716-0916-3_14, © Springer Science+Business Media, LLC, part of Springer Nature 2021

ECs are the cells that line the blood vessel lumens and are, of course, central to the process. Our group has tested EC sources from several species (namely, human, mouse, rabbit, and sheep) [3, 5, 9] and tissues of origin, including ECs from the umbilical cord or peripheral blood (i.e., endothelial colony-forming cells or ECFCs) [3], tissue-resident ECs (e.g., from adipose and dermal tissues) [9], umbilical vein (human umbilical vein endothelial cells or HUVECs) [3], and induced pluripotent stem cell (iPSC)-derived ECs [11]. We have consistently validated the ability of all these types of ECs to form perfused vessels using the in vivo vascular network forming assay [12].

The second cell type required for the formation of vascular networks is perivascular cells. Sources of perivascular cells include smooth muscle cells (SMCs) [3], mesenchymal stem cells (MSCs) [5, 7], and fibroblasts [13]. As with ECs, investigators have routinely tested multiple types of supporting perivascular cells using the in vivo vascular network forming assay [12], and there is consensus in that perivascular cells are critical to facilitating the engraftment of ECs [5].

The in vivo vascular network forming assay is very versatile, and its applications for both basic vascular biology and translational studies are numerous [12]. For example, using this assay, one could quickly characterize new sources of ECs such as those derived from embryonic stem cells, ECs isolated from pathological specimens, and ECs genetically modified via gene editing. Investigators could also histologically evaluate the characteristics of the newly formed blood vessels, including their size, morphology, microvascular density, whether the vessels are perfused or not. Also, applying routine immunofluorescence, we can identify the implanted ECs (e.g., using species-specific antibodies against EC markers) and perivascular cells (e.g., by staining for a-SMA), and distinguish them from the host cells [14]. We can also analyze and evaluate the functionality of the vessels, including the presence of proliferative cells (e.g., Ki67-positive cells) and apoptosis (TUNEL assay) [15–17]. The in vivo vascular network forming assay also allows testing pharmaceutical enhancers or inhibitors of blood vessel formation, which could be screened by simply adding the candidate molecules into the hydrogel mixture. Moreover, the entire process of vascular network formation occurs in a short period (typically, perfused vessels are formed within 7 days after implantation), thus facilitating the studies [17]. In summary, the in vivo vascular network forming assay is ideally suited for studies on the cellular and molecular mechanisms of vascular network formation and for developing strategies to vascularize tissues.

In this chapter, we provide an updated version of the in vivo vascular network forming assay. Of note, we describe an alternative hydrogel to Matrigel. Originally, this assay was developed with Matrigel due to its convenience to obtain and handle. However,

Matrigel is not fully defined chemically and thus has some drawbacks. Indeed, Matrigel is a mixture of extracellular matrix components isolated from Engelbreth-Holm–Swarm mouse sarcomas and contains various undefined proteins [2]. Thus, one should always be cautious when interpreting experiments on cellular activities in Matrigel. Moreover, there are significant limitations to Matrigel when considering translational studies, including its tumorigenic origin, diverse composition, and batch-to-batch variability in terms of mechanical and biochemical properties. Over the last decade, our group has characterized several alternative options to Matrigel (both synthetic and natural sources), and we have found that human vascular cells can form robust vascular networks in multiple types of hydrogels [4, 14–16, 18, 19]. Here, we describe a collagen–fibrin hydrogel mixture that is easy to use, produces very reproducible in vivo outcomes, and has a well-defined composition. Indeed, the assay we describe herein entails the injection of a collagen–fibrin hydrogel mix containing ECFCs and MSCs, which supports the formation of a robust vascular network within 7 days [19]. The assay is simple to perform, and it does not require any incision or surgical procedure, which reduces the potential influence of wound healing. Lastly, the assay is compatible with many immunodeficient mouse backgrounds, including athymic nude, SCID, and NSG mice.

2 Materials

Prepare all solutions using PBS or distilled water and analytical grade reagents. Prepare and store all reagents at 4 °C temperature (unless indicated otherwise). Diligently follow all waste disposal regulations when disposing of waste materials.

2.1 Cell Culture of Human ECFCs and MSCs

1. We used human cord blood–derived ECFCs and adipose-derived MSCs as the standard cell sources. It is recommended to include ECFC/MSC as a positive control in all experiments. The isolation, characterization, and cultivation of ECFCs and MSCs are out of the scope of this chapter, but detailed procedures can be found in our previous publications [12]. Alternatively, we and others have also confirmed that HUVECs can be used instead of ECFCs [3]. Also, purified ECs can be purchased from several commercial sources like STEMCELL Technologies Inc or Lonza.

2. EC culture medium: Endothelial cell growth medium 2 (EGM2, except for hydrocortisone; PromoCell, Cat# C22111) supplemented with 10% FBS and 1× glutamine–penicillin–streptomycin.

3. MSC culture medium: Mesenchymal stem cell growth medium (MSCGM; ATCC, Cat# PCS-500-030) supplemented with 10% FBS and 1× glutamine–penicillin–streptomycin.

2.2 Collagen–Fibrin Hydrogel

1. Concentrated bovine collagen solution: We validated Cultrex® Bovine Collagen I (protein concentration, 5 mg/mL) from Trevigen. Similar products isolated from rat tail or porcine skin should also work, but the stock concentration has to be higher than 5 mg/mL. Keep collagen solution at 4 °C and handle it on ice.

2. Bovine fibrinogen solution: Prepare freshly 30 mg/mL of fibrinogen concentrated solution by dissolving bovine fibrinogen (Sigma-Aldrich, Cat# F8630) in prewarmed 0.9% sodium chloride saline. Keep the solution in a 37 °C water bath with occasional vortexing until dissolved.

3. Store components for collagen–fibrin hydrogels, including 1 M HEPES (4-(2-hydroxyethyl)-1-piperazineethanesulfonic acid) buffer, 10× M199 or DMEM media, fetal bovine serum (FBS), and 1 mg/mL of human fibronectin (hFN), at 4 °C and handle them on ice.

4. Final composition of collagen–fibrin hydrogel includes collagen (3 mg/mL), human fibronectin (30 μg/mL), FBS (10% v/v), HEPES (25 mM), and fibrinogen (3 mg/mL) at pH 7.4. Use Table 1 to calculate the amount of collagen–fibrin gel needed for the experiment and follow Subheading 3.1 to prepare the hydrogel solution.

5. Thrombin solution: Dissolve 50 U/mL of thrombin from bovine plasma (Sigma-Aldrich, Cat# T4648) in PBS. Aliquot and store at −20 °C.

2.3 Immuno-fluorescent Staining

1. Tris–EDTA antigen retrieval buffer: Add 0.6 g of Tris-base (Trizma® base from Sigma-Aldrich), 1 mL of 0.5 M EDTA, 250 μL of Tween 20 in 500 mL of ddH$_2$O. Mix and adjust pH with HCl to 9.0. Heat to 90–95 °C on a magnetic stir plate with gentle mixing.

2. 5% serum blocking buffer: Add 0.5 mL of normal horse serum (Vector Laboratories) in 9.5 mL of PBS. Mix and use within 6 h.

3 Methods

Prior to the experiment, make sure there are enough ECFCs and MSCs in culture; 0.8×10^6 ECFCs and 1.2×10^6 MSCs will be required for each implant and mouse. We suggest preparing 1.25 times more amount of cell–hydrogel mixture to compensate for the

Table 1
Composition of the collagen–fibrin hydrogel

Stock Conc.		Final Conc.
Collagen (5 mg/mL)	600 μL	3 mg/mL
10× M199	100 μL	1×
hFN (1 mg/mL)	30 μL	30 μg/mL
FBS	100 μL	10%
HEPES (1 M)	25 μL	25 mM
NaOH (1N)	~15 μL	pH 7.4 (by color)
Fibrinogen (30 mg/mL)	100 μL	3 mg/mL
ddH$_2$O	30 μL	
Total	1000 μL	

potential loss during handling and the dead volume of the syringe. For example, we would prepare 1-mL of a cell–hydrogel mixture containing 4×10^6 ECFCs and 6×10^6 MSCs but would only inject four mice, each with a 200-μL plug. Carry out all procedures in a tissue culture hood with sterile equipment.

3.1 Preparation of Collagen–Fibrin Hydrogel

1. Calculate the amount of collagen–fibrin gel needed for the experiment (Table 1). Keep all reagents on ice. Precool 1000 μL pipette tips at −20 °C for collagen solution handling.

2. Prepare the fibrinogen solution freshly by dissolving bovine fibrinogen (30 mg/mL) in 0.9% sodium chloride saline. Keep the vial in a 37 °C water batch until all fibrinogen is dissolved with occasional vortexing.

3. Under a tissue culture hood, carefully mix the reagents in a suitable vial on ice following this order: ddH$_2$O, 10× M199 medium, 1 M HEPES, and 5 mg/mL collagen solution. Use precooled pipette tips to transfer collagen solution and mix the hydrogel. Pay attention to the color change of pH indicator phenol red in the M199 medium. While adding 1 M HEPES, the color should turn to red for neutral pH (6.8–7.4). After mixing the collagen solution, the pH should decrease to ~pH 2 and show a yellow color. Then, slowly titrate the pH value back to neutral by adding 1N NaOH. Mix the solution thoroughly by gentle pipetting (avoiding bubbles) for accurate pH measurement. Transfer a tiny amount (~1 μL) of hydrogel solution to a pH strip (PH Test Strips, 0-14 PH, EMD Millipore cat# 9590) to confirm a neutral pH.

4. Once you obtain a neutral hydrogel solution, then add fibronectin, FBS, and fibrinogen to the solution and mix thoroughly. FBS and fibronectin are optional and can be replaced by the same amount of ddH$_2$O. Keep the hydrogel solution on ice and use it within 1 h (*see* **Note 1**).

3.2 Preparation of Cell–Hydrogel Mixture for Injection

1. Aspirate the medium of each cell culture plate and wash the ECFCs and MSCs with 10 mL of PBS. Remove PBS and add 5 mL of the trypsin–EDTA solution to each 100-mm plate. Gently shake the plates to evenly distribute the trypsin–EDTA solution. Incubate the plate for 4–5 min. Gently tap the plate to see the detached cells in suspension under an inverted microscope.

2. When cells completely detach from the plate, add 5 mL of DMEM/10% FBS, and collect the cell solution into a 15-mL conical tube. Take 10 μL of the solution to count the cells in a hemocytometer and work out the total number of ECFCs and MSCs harvested, respectively.

3. Transfer 4×10^6 ECFCs ($5 \times 0.8 \times 10^6$ cells) and 6×10^6 MSCs ($5 \times 1.2 \times 10^6$ cells) together into a single 50-mL conical tube. This is the total amount of cells required for four implants with some extra amount to account for losses during handling. Centrifuge the cells at 1200 rpm ($290 \times$ g) and remove the supernatant (*see* **Note 2**).

4. Resuspend the cell pellet in 1 mL of ice-cold collagen–fibrin hydrogel solution. Mix the cells very gently to avoid any bubbles within the mixture. Keep the cell–hydrogel mixture on ice and inject it within 30 min (*see* **Note 3**).

3.3 Injection into Immunodeficient Nude Mice

1. All animal experiments are carried out with 6-week-old athymic nude mice. Mice are housed in compliance with Boston Children's Hospital guidelines, and all animal-related protocols are approved by the Institutional Animal Care and Use Committee (*see* **Note 4**).

2. Prior to the injection, anesthetize the mice by placing them into a gas chamber delivering isoflurane. Allow the mice to inhale the isoflurane for approximately 2 min until they are unresponsive to toe pinch and monitor their respiration by inspection (*see* **Note 5**).

3. Gently mix the cell–hydrogel mixture again and immediately load the mixture into 1-mL sterile syringes, and place 26-gauge needles with their caps on the tips of the syringes. Keep the loaded syringes horizontally on ice to prevent ununiform cell distribution (*see* **Note 6**).

4. Load a separate 1-mL sterile syringe with the thrombin solution and place a 30-gauge needle with its cap on the tip of the syringe.

5. Disinfect the injection area of nude mouse skin with ethanol pads. For each mouse, inject 50-µL of thrombin solution first and then, at the same location, 200-µL of the cell–hydrogel solution subcutaneously into the upper dorsal region. The fibrinogen portion in the hydrogel solution will be enzymatically converted by thrombin to fibrin within minutes to form a gel plug, which will become a small but appreciable bump under the skin (*see* **Note 7**).

6. After the injection, place the mice on a layer of gauze for comfort and warmth and observe them until they become ambulatory. Then, observe the mice daily for the first 3 days.

3.4 Harvesting the Implants

1. One week after the implantation, euthanize the mice by placing them in a gas chamber, delivering compressed CO_2 gas. The collagen–fibrin plugs should be appreciable under the skin (*see* **Note 8**).

2. Cut open the skin near the area of the original injection and surgically removed the collagen–fibrin plug. Digital photographs of the retrieved plugs with a scale are advised.

3. Place the harvested collage–fibrin plugs into histological cassettes and fix them in 10% neutral buffered formalin under a chemical hood overnight at room temperature.

4. After fixation, wash out the 10% neutral buffered formalin with ddH_2O several times and place the histological cassettes at 4 °C in 70% ethanol until histological evaluation.

3.5 Evaluation of Vascular Networks in Explanted Collagen–Fibrin Plugs

1. For histological evaluation, explanted collagen–fibrin plugs are embedded in paraffin and sectioned (7 µm-thick sections) using standard procedures.

2. Carry out Hematoxylin and Eosin (H&E) staining of the explant sections following standard protocols.

3. Quantify microvessel density (MVD) by evaluation of 10–30 randomly selected fields (0.1 mm^2 each; 40× objective lens) of H&E stained sections taken from the middle part of the implants. Microvessels can be identified as luminal structures containing red blood cells and counted. Calculate MVD as the average number of red blood cell-filled microvessels from the fields analyzed and expressed as vessels per square millimeter in the image (*see* **Notes 9** and **10**).

3.6 Evaluation of Human Lumens and Perivascular Coverage by Immunofluorescent Staining

1. Deparaffinize and rehydrate the paraffin-embedded sections by sequential immersion in xylene and then 100%, 90%, 80%, and 50% ethanol for 5 min in each step. Rinse the sections in PBS.

2. Heat the sections in Tris–EDTA antigen retrieval buffer (the default antigen retrieval buffer for most the antibodies we used unless mentioned) to 90–95 °C for 30 min. Rinse the sections in PBS.

3. Block the sections for 30 min in blocking solution (5% normal horse serum in PBS).

4. Incubate the sections with primary antibody solution (in 5% blocking serum) for 1 h at room temperature or 4 °C overnight. Validated primary antibodies include mouse anti-human CD31 (human EC-specific; Agilent Technologies Inc, Clone JC70A; 1:50 dilution); mouse anti-human vimentin (human EC and MSC-specific; Abcam, Clone V9; 1:100 dilution); mouse anti-α-smooth muscle actin (α-SMA; reactive with both human and mouse; Sigma-Aldrich; Clone 1A4; 1:100 dilution); and rabbit anti-α-smooth muscle actin (α-SMA; reactive with both human and mouse; Abcam; Clone ab5694; 1:100 dilution). Afterward, wash the sections with PBS twice (*see* **Note 11**).

5. Incubate the sections with fluorophore-conjugated secondary antibody solution (1:200 in 5% blocking serum) for 1 h at room temperature. Validated secondary antibodies include horse anti-mouse IgG conjugated with fluorescein (FI-2000) or Texas Red (TI-2000) and Texas Red goat anti-rabbit IgG (TI-1000; all from Vector Laboratories). Afterward, wash the sections with PBS twice. Protect the slides from light after adding secondary antibodies to avoid photobleaching.

6. Counterstain cell nuclei with DAPI solution at room temperature for 10 min. Wash the sections with PBS twice.

7. Wash the slides with ddH$_2$O once and mount them with ProLong antifade mounting medium (Thermo Fisher Scientific). Keep slides at 4 °C and protect them from light until imaging. Obtain high-resolution images with a confocal microscope (e.g., Leica TCS SP2 confocal system), using a 63× objective lens.

3.7 Evaluation of Cell Proliferation and Apoptosis in Vascular Networks

1. Evaluation of cell proliferation can be done by immunofluorescent staining of Ki67 on the paraffin-embedded sections. We have validated the rabbit anti-Ki67 antibody (reactive with both human and mouse; Abcam; Clone ab15580; 1:100 dilution) by the same staining protocol of Subheading 3.6. Double staining with human EC-specific CD31 may be required to distinguish proliferating human and mouse cells.

2. Evaluation of cell apoptosis can be done by using Click-iT™ Plus TUNEL Assay for In Situ Apoptosis Detection, Alexa Fluor™ 488 dye (Thermo Fisher Scientific), or other similar TUNEL assays. After performing the TUNEL assay according to the manufacturer's protocol, the sections can be stained by immunofluorescence with human EC-specific CD31 antibody using the same staining protocol of Subheading 3.6 to distinguish the origin of apoptotic cells.

4   Notes

1. The composition of collagen–fibrin hydrogel gel is adjustable. For example, FBS and fibronectin are optional and can be replaced by the same amount of ddH$_2$O. Additional reagents, including growth factors, cytokines, blocking antibodies, ECM proteins, concentrated conditioned medium, cell extract, or small pharmaceutical molecules, can be added into the hydrogel mixture to test theirs in vivo effects. Any protein or peptide should be added after adjusting the pH value to natural to prevent inactivation due to acidic denaturation.

2. The total numbers of cells and the ratio between ECs and MSCs can be adjusted according to experimental designs. We have validated the successful formation of vascular networks by implanting total cell numbers ranged from 1×10^6 to 6×10^6 per 200-µL plug, with EC-to-MSC ratios from 4:1 to 1:4.

3. The time between resuspension of the cells in collage-fibrin hydrogel and the subcutaneous injection into the mice should be kept to a minimum to maintain cell viability. We recommend the preparation of the hydrogel solution prior to harvesting the cells.

4. We have confirmed comparable levels of vascular network formation by injecting ECFCs and MSCs in different immunodeficient murine backgrounds, including athymic nude (nomenclature: Crl:NU(NCr)-$Foxn1^{nu}$), BALB/c nude (CAnN.Cg-$Foxn1^{nu}$/Crl), CD-1 nude (Crl:CD1-$Foxn1^{nu}$), NOD.SCID (NOD.CB17-$Prkdc^{scid}$/J), and NSG (NOD.Cg-$Prkdc^{scid}$ $Il2rg^{tm1Wjl}$/SzJ) mice.

5. Other analgesics are compatible with this assay. However, due to the low invasiveness and the quick nature of this procedure, analgesics are usually not required. Consult with your animal housing facility to find the best practice for your experiment.

6. Injecting cells through needles smaller than 26-gauge may damage cell viability.

7. The purpose of the preinjected thrombin solution is to convert the fibrinogen into fibrin plug immediately after injecting it into the subcutaneous space. The quick formation of fibrin gel significantly improves cell retention and the proper volume of the hydrogel implant. The effect of thrombin is transient. Without adding thrombin, the hydrogel can still solidify by the formation of a collagen matrix at body temperature. However, the process may take 30 min or longer.

8. We routinely exam the formation of vascular networks after 7 days. The harvesting time points can be adjusted due to the purpose of the research. We have observed the earliest perfused

blood vessels at day 3 and reliable vascularization at day 5. The vascular networks are stable for at least 4 weeks. However, the Matrigel may be more suitable for long-term grafting due to the quick degradation of collagen–fibrin hydrogel.

9. The microvessel density formed by ECFCs and MSCs after 7 days should range from 80 to 250 vessels/mm^2. We recommend having at least four plugs per group and repeat 3 times to reduce the variation between individual mice.

10. Several studies using similar in vivo vascular network forming assays have shown the dysfunction of blood vessels by implanting pathological vascular cells. Some of the features are obvious while examining the H&E images and recapitulate their clinical pathology. For example, implanting cells isolated from the infantile hemangioma displayed unusually high microvessel density [20, 21]. Tie2-mutated HUVECs formed enlarged lumens that recapitulate the venous malformation lesions [22].

11. Human-specific ECs can also be stained by *Ulex europaeus* agglutinin I (UEA-I), a lectin that binds with high affinity to human (but not mouse) ECs. After performing antigen retrieval, incubate the sections with fluorescein- or rhodamine-conjugated UEA-I lectin (Vector Laboratories; 20 µg/mL in PBS containing 1 mM Calcium and 1 mM Magnesium ions) at room temperature for 30 min. Rinse the section with PBS twice. UEA-I lectin binding method is compatible with antibody-based staining and useful to avoid the cross-reactions of secondary antibodies.

Acknowledgments

This work was supported by the National Institutes of Health grants R01AR069038 and R01HL128452 to J. M.-M.

References

1. Kastana P, Zahra FT, Ntenekou D et al (2019) Matrigel plug assay for in vivo evaluation of angiogenesis. Methods Mol Biol 1952:219–232

2. Benton G, Arnaoutova I, George J et al (2014) Matrigel: from discovery and ECM mimicry to assays and models for cancer research. Adv Drug Deliv Rev 79-80:3–18

3. Melero-Martin JM, Khan ZA, Picard A et al (2007) In vivo vasculogenic potential of human blood-derived endothelial progenitor cells. Blood 109:4761–4768

4. Melero-Martin JM, Bischoff J (2008) Chapter 13. An in vivo experimental model for postnatal vasculogenesis. Methods Enzymol 445:303–329

5. Melero-Martin JM, De Obaldia ME, Kang SY et al (2008) Engineering robust and functional vascular networks in vivo with human adult and cord blood-derived progenitor cells. Circ Res 103:194–202

6. Traktuev DO, Prater DN, Merfeld-Clauss S et al (2009) Robust functional vascular network formation in vivo by cooperation of adipose progenitor and endothelial cells. Circ Res 104:1410–1420

7. Au P, Daheron LM, Duda DG et al (2008) Differential in vivo potential of endothelial

progenitor cells from human umbilical cord blood and adult peripheral blood to form functional long-lasting vessels. Blood 111:1302–1305

8. Risau W, Flamme I (1995) Vasculogenesis. Annu Rev Cell Dev Biol 11:73–91

9. Lin RZ, Moreno-Luna R, Munoz-Hernandez R et al (2013) Human white adipose tissue vasculature contains endothelial colony-forming cells with robust in vivo vasculogenic potential. Angiogenesis 16:735–744

10. Wang ZZ, Au P, Chen T et al (2007) Endothelial cells derived from human embryonic stem cells form durable blood vessels in vivo. Nat Biotechnol 25:317–318

11. Neumeyer J, Lin RZ, Wang K et al (2019) Bioengineering hemophilia A-specific microvascular grafts for delivery of full-length factor VIII into the bloodstream. Blood Adv 3:4166–4176

12. Wang K, Lin RZ, Melero-Martin JM (2019) Bioengineering human vascular networks: trends and directions in endothelial and perivascular cell sources. Cell Mol Life Sci 76:421–439

13. Costa-Almeida R, Gomez-Lazaro M, Ramalho C et al (2015) Fibroblast-endothelial partners for vascularization strategies in tissue engineering. Tissue Eng Part A 21:1055–1065

14. Lin RZ, Melero-Martin JM (2012) Fibroblast growth factor-2 facilitates rapid anastomosis formation between bioengineered human vascular networks and living vasculature. Methods 56:440–451

15. Chen YC, Lin RZ, Qi H et al (2012) Functional human vascular network generated in photocrosslinkable gelatin methacrylate hydrogels. Adv Funct Mater 22:2027–2039

16. Lin RZ, Chen YC, Moreno-Luna R et al (2013) Transdermal regulation of vascular network bioengineering using a photopolymerizable methacrylated gelatin hydrogel. Biomaterials 34:6785–6796

17. Lin RZ, Moreno-Luna R, Li D et al (2014) Human endothelial colony-forming cells serve as trophic mediators for mesenchymal stem cell engraftment via paracrine signaling. Proc Natl Acad Sci U S A 111:10137–10142

18. Allen P, Melero-Martin J, Bischoff J (2011) Type I collagen, fibrin and PuraMatrix matrices provide permissive environments for human endothelial and mesenchymal progenitor cells to form neovascular networks. J Tissue Eng Regen Med 5:e74–e86

19. Lin RZ, Melero-Martin JM (2011) Bioengineering human microvascular networks in immunodeficient mice. J Vis Exp:e3065

20. Greenberger S, Boscolo E, Adini I et al (2010) Corticosteroid suppression of VEGF-A in infantile hemangioma-derived stem cells. N Engl J Med 362:1005–1013

21. Greenberger S, Yuan S, Walsh LA et al (2011) Rapamycin suppresses self-renewal and vasculogenic potential of stem cells isolated from infantile hemangioma. J Invest Dermatol 131:2467–2476

22. Boscolo E, Limaye N, Huang L et al (2015) Rapamycin improves TIE2-mutated venous malformation in murine model and human subjects. J Clin Invest 125:3491–3504

## Chapter 15

Assessment of Vascular Patterning in the Zebrafish

Amber N. Stratman and Brant M. Weinstein

Abstract

The zebrafish has emerged as a valuable and important model organism for studying vascular development and vascular biology. Here, we discuss some of the approaches used to study vessels in fish, including loss-of-function tools such as morpholinos and genetic mutants, along with methods and considerations for assessing vascular phenotypes. We also provide detailed protocols for methods used for vital imaging of the zebrafish vasculature, including microangiography and long-term time-lapse imaging. The methods we describe, and the considerations we suggest using for assessing phenotypes observed using these methods, will help ensure reliable, valid conclusions when assessing vascular phenotypes following genetic or experimental manipulation of zebrafish.

Key words Zebrafish, Vascular biology, Vasculature, Blood vessel, Lymphatic vessel, Endothelium, Morpholino, Microangiography

1 Introduction

The zebrafish continues to gain traction as a valuable vertebrate model system for studying vascular development and angiogenesis in vivo. An extensive number of studies now demonstrate the conservation of vascular development between zebrafish, mice, and humans in terms of both anatomy and function. Benefits of the zebrafish compared to other vertebrate model systems include (1) genetically accessible with externally fertilized, optically clear embryos and larvae that permit high-resolution imaging of developing tissues; (2) large clutch sizes for experimentation and genetic screens; and (3) embryos that can survive and continue to develop for several days without a functional circulation via passive diffusion of oxygen and waste across the skin. Although the beneficial experimental attributes of zebrafish have made this an extremely appealing and approachable model system for many labs, like any model organism there are limitations to the studies that can be performed and their interpretation. In this chapter, we will cover loss-of-function approaches and imaging techniques available to study

Domenico Ribatti (ed.), *Vascular Morphogenesis: Methods and Protocols*, Methods in Molecular Biology, vol. 2206, https://doi.org/10.1007/978-1-0716-0916-3_15, © Springer Science+Business Media, LLC, part of Springer Nature 2021

vascular development and vascular patterning in the zebrafish, including some of the pitfalls in interpreting data generated using currently available approaches.

1.1 Morpholinos

Prior to the development of genome editing technologies, morpholinos (MOs) were the only available, reasonably well-validated targeted loss-of-function method in the zebrafish. Morpholinos are specifically modified stable antisense oligomers designed to block either gene translation or gene splicing when injected into embryos [1], and absent alternative reverse genetic tools, morpholino-based experiments rapidly gained popularity in the fish community. While revolutionary for the field at the time, since their advent concerns have been raised about the specificity of phenotypes generated from MO injections, particularly the potential for off-target and/or pleiotropic effects [1–4]. These concerns have been heightened by growing evidence that morpholinos and genetic mutants for the same genes often do not yield comparable phenotypes [2]. In order to promote greater validity and rigor of experimental studies using morpholinos, attempts have been made to develop standards and "community guidelines" for their use. One recently circulated set of recommendations suggests use of (1) multiple MO targets, (2) dose response curves for titration of the MO, (3) RNA/DNA rescue experiments, and importantly, (4) validation with a genetic mutant where possible [1]. Further, it is expected that proper morpholino use requires rigorous validation of their specific blocking ability, either by assessing loss of protein production for translation blocking MOs (when specific and/or species cross-reactive antibodies are available) or by assessing the generation of alternate splice variants by RT-PCR when using splice blocking MOs. MO based experiments should also be "rescued" by coadministration of mRNA globally or more ideally by administration of cell-type specific transgenes expressing a wild type copy of the targeted gene of interest, to strengthen arguments for cell autonomous versus nonautonomous effects. However, the best accepted validation for any MO based observations is confirmation of the same phenotype(s) in a zebrafish genetic mutant. As discussed further below, many mutants are now available from forward-genetic ENU mutagenesis screens, "tilling" approaches, or genome editing methods. CRISPR editing technologies have made it easy and straightforward for any lab to carry out reverse genetic mutation of virtually any gene of interest [1, 4–28]. In short, when possible, all MO experiments should be confirmed and validated with a genetic mutant before they are used extensively for experimental studies.

Common off-target effects noted in morpholino-injected animals include p53-mediated cell death, defective circulation and "ballooning" (edema), and developmental delays/stunting of embryos and larvae. Altered vascular development and patterning should be interpreted with extreme caution in the presence of any of these phenotypes, as vessel growth and patterning in developing embryos and larvae is highly dependent on the integrity of adjacent tissues and of the animal as a whole (see below for further detailed discussion). The recently published community guidelines noted above include recommendations on how to interpret MO phenotypes, how to validate these phenotypes, and current expectations in the field for MO usage [1].

Although MO approaches, and especially their use without proper validation and controls, have justifiably come under a great deal of skepticism in recent years due to well-documented off-target effects and lack of correlation with mutant phenotypes, it is worth noting a few positive/beneficial features of MOs that make their continued use (with caution) worthwhile. First, translation-blocking morpholinos can block both zygotic and maternally supplied forms of target genes, conversely maternal-zygotic genetic mutants can be difficult or impossible to obtain. Second, MOs can be used rapidly on any zebrafish line or strain, unlike genetic mutants, which can take months to cross onto appropriate genetic backgrounds (when using combinations of mutants and transgenes, this can take many generations). Third, multiple morpholinos can be injected simultaneously to target two, three, or more members of a gene family with potentially overlapping functions, or designed to target consensus domains in gene families, such as with redundant transcription factors, to impair function of the whole gene family with a single reagent. Finally, recent work has shown that upregulation of related compensating gene family members can sometimes occur in genetic mutants, likely by triggered mutant mRNA decay, while this does not appear to take place in morpholino-injected animals [3, 29], arguably making morpholinos a better representation of targeted loss of gene function in some cases. However, if used extremely extensive additional work is needed to validate all interpretation and distinguish phenotypes from off-target effects in any MO based study.

1.2 Genetic Mutants

One of the major strengths of the zebrafish is the ability to carry out large-scale, forward-genetic phenotype-based ENU mutagenesis screens for unbiased identification of mutants effecting developmental processes. Large numbers of mutants have now been isolated affecting almost every conceivable developmental program, leading to innumerable new discoveries in the vascular field and beyond [30–37]. Forward genetic screens have been an extremely powerful approach for identification of genes necessary for developmental processes during times when reverse-genetic approaches for targeting specific genes of interest were not available in the zebrafish. "Tilling" approaches of mutagenized fish have been used to screen for ENU-induced mutations in specific genes by capitalizing on high-throughput next generation sequencing of libraries [38]; however, these approaches remain laborious, expensive, and fail to recover some genes of interest as they are not easily mutagenized. More recently, genome editing technologies, and particularly CRISPR/Cas9 and Cpf1/Cas12a-based methods, have made reverse-genetic targeting of virtually any gene of interest easy and cost effective for most labs [5–7, 26–28]. A number of resources now make identifying "ideal" CRISPR cut sites, primer selection, and mutation screening relatively straight forward [5, 39–43]. Generating specific alleles by homologous integration is still challenging in the zebrafish, but the technology and resources to generate "knockin" mutants and floxed genes using CRISPR methods is also rapidly improving and is likely to soon become accessible to most zebrafish labs [7, 26, 41].

Although genetic mutants are an effective approach for loss-of-function analysis, and those generated via CRISPR relatively straight forward to generate, the resulting phenotypes must still be interpreted rigorously. While potentially resulting from defects in the targeted gene of interest, vascular phenotypes observed in animals that display significant cell death, defective circulation, edema, and developmental delay/stunting of embryos and larvae may also represent secondary, indirect consequences of the genetic mutation and may not reflect a direct requirement for the targeted gene during processes of vascular development. It is also worth noting that transient CRISPR approaches involving injection of guide RNAs into animals transgenically expressing Cas9 in specific tissues ("CRISPRi") or primary analysis of nonstable, G0 CRISPR injections are subject to the same general concerns regarding off-target effects as morpholinos, although the spectrum of off-target effects observed with the two approaches could be distinct. Phenotypes observed in either morpholino- or CRISPR-injected animals must be interpreted with caution, and ideally should be verified using a stably transmitted germline genetic mutation.

The ease with which CRISPR mutants can be generated also makes it possible to simultaneously induce and screen for mutations in multiple genes at the same time. This is advantageous for studying the role of gene families in the zebrafish. Gene paralogs are more common in zebrafish than in mammals due to a genome duplication that occurred during the evolution of teleost fish, making genetic compensation by alternate gene family members and obscured phenotypes quite common in zebrafish genetic mutants [3, 29]. Therefore, it has become more common to see reports in which gene promoters, whole exons/targeted functional domains, or double and even triple loss of function mutants to closely related gene family members have been generated to observe actionable phenotypes; in some cases phenotypes comparable to those noted in morpholino-injected animals have been unmasked [3, 29, 44].

1.3 Assessing Vascular Phenotypes in Loss-of-Function Models

Whether morpholinos or mutants are used, assessment of vascular phenotypes involves a number of specialized considerations, as proper vessel formation can be disrupted by generic developmental delays, global and local changes in blood flow, and local or generalized defects in nonvascular tissues. Below, we discuss some points to consider in assessing vascular phenotypes and discuss potential pitfalls in utilizing and understanding zebrafish vascular development data.

1.3.1 Where Is the Gene Expressed?

An early consideration in assessing the role of a particular gene in vascular function should be understanding where the gene is expressed—regardless of the gene manipulation approach chosen for downstream analysis. This is generally assessed using whole mount in situ hybridization (WISH) of zebrafish embryos and early larvae [45], or in situ hybridization of tissue sections in older animals. When cross reactive antibodies are available, immunostaining analysis can also be carried out to determine protein localization either on whole mount embryos or on tissue sections. If a gene shows a restricted vascular expression pattern (Fig. 1a, b), then it is likely to have a vascular-associated function and/or vascular autonomous-phenotype related specifically to the gene. Conversely, if the gene shows an exclusively nonvascular expression pattern (Fig. 1c, d), it is not reasonable to assume that the gene has a vascular-autonomous or restricted function, and any vascular phenotypes that are observed could be indirect or secondary to the function of the targeted gene. For many genes however, expression is observed in both vessels and in nonvascular tissues (Fig. 1e), and some common-sense judgement must be applied in assessing whether observed loss-of-function phenotypes make sense in terms of the expression pattern of the gene. If significant nonvascular expression is observed, explicitly designed experimental approaches should be utilized to assess the cell autonomy of gene function, including transplantation experiments or tissue-specific/cell type specific transgenic expression to "rescue" the phenotype (Fig. 1f–h).

1.3.2 Assessing General Morphology and Development in Morphants or Mutants

Vascular phenotypes that seem striking can often appear as a consequence of secondary, nonspecific developmental delays or general changes in nonvascular tissues or organs. Staging tables are available for the zebrafish [18] and it is important to determine whether morpholino injections or genetic mutants result in either overall delays to development or gross embryonic/larval tissue patterning, or localized defects in the development and morphology of specific tissues or organs (Fig. 2a, b). For morphants, this should be done by comparing animals injected with a specific MO to siblings injected with a control MO and uninjected siblings. For genetic mutants, phenotypes should be compared between homozygous mutants and wild-type/heterozygous siblings from the same clutch

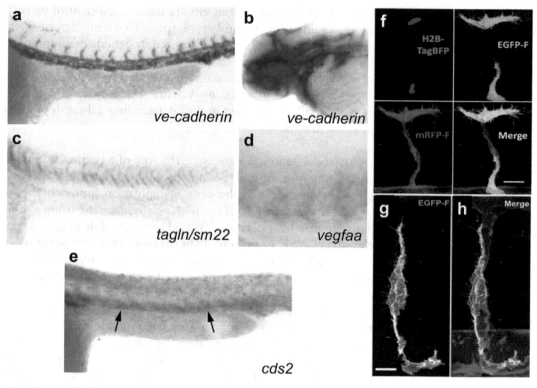

Fig. 1 Assessing EC autonomous gene function. (**a–e**) Whole mount in situ hybridization of 24–48 hpf zebrafish embryos. (**a, b**) *ve-cadherin* labeling of the vasculature in the trunk (**a**) and head (**b**) of a zebrafish embryo showing vascular specific labeling. (**c, d**) *tagln/sm22* (**c**) and *vegfaa* [36] (**d**) showing labeling of nonvascular tissues. (**e**) *cds2* [37] labeling of both vascular (arrows) and nonvascular tissues. (**f**) Confocal micrograph of a growing trunk ISV segment at 32 hpf in a *Tg(kdrl:mRFP-F)^{y286}* embryo, mosaically expressing *Tol2(fli1a:H2B-TagBFP-p2A-egfp-F)* transgene, showing blue, green and red fluorescent channels and all three merged [46]. Scale bar: 20 μm. (**g, h**) High-magnification images of GFP fluorescence (**g**) and merged GFP/BFP/RFP fluorescence (**h**) images of *Tol2(fli1a:H2B-TagBFP-p2A-egfp-F)* mosaic-expressing EC in a trunk ISV segment at 42 hpf in a *Tg(kdrl:mRFP-F)^{y286}* transgenic background embryo [46]. Scale bar: 10 μm

of eggs. In either case, if the overall development of the animal is significantly delayed or readily apparent gross morphological changes are noted compared to the control sibling animals, vascular phenotypes should be interpreted with caution. For example, reduced head size and/or extensive cell death in the central nervous system are common off-target effects noted in both mutant and morphant models, and this secondarily results in vessel defects in the heads of affected animals (Fig. 2g, h). The timing and extent of trunk intersegmental vessel growth is frequently used as a convenient and quantitatively robust assay for assessing angiogenesis defects in zebrafish embryos, but delayed development or abnormal formation of the adjacent somites results in defects in vessel sprouting and growth that, again, are not directly linked to vascular-specific functions (Fig. 2c–f, i, j). In fact, completely wild type

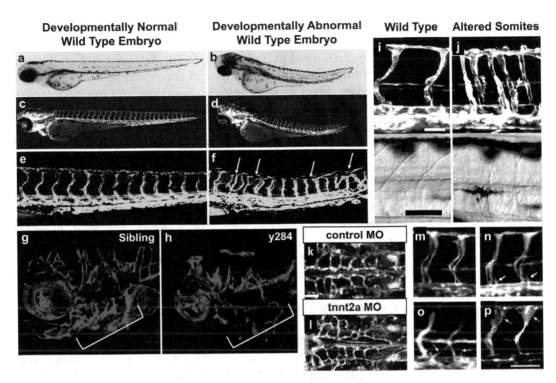

Fig. 2 Assessing embryonic morphology and effects on vascular patterning. (**a, c, e**) Images of a 3 dpf wild type zebrafish embryo that is morphologically normal (**a**), demonstrating normal patterning of the trunk intersegmental vessels (**c, e**). (**b, d, f**) Images of a 3 dpf zebrafish embryo that is developmentally abnormal, despite deriving from a completely wild type zebrafish population, demonstrating striking developmental delay (**b**), marked gross morphologic defects (**b**), and mispatterning of the intersegmental vessels (arrows—**f**). (**g, h**) Mispatterning and missing vasculature (brackets) in the head of y284 mutants that have small heads and eyes. Caution must be taken to understand if the vascular defects noted are primary or solely a consequence of the head being overall smaller. (**i, j**) Confocal images of intersegmental vessels in 48 hpf *Tg(fli1:egfp)y1* wild-type (**i**) and y66 mutant (**j**) animals. Images in (**b–i**) are all oriented with anterior to the left and dorsal up. Images demonstrate improper somite formation (lack of chevron shaped somite segments as shown in the bright field images) can alter the patterning of the vasculature [34]. Scale bars: 50 µm. (**k–p**) Analysis of the vasculature in the absence of blood flow. (**k, l**) Images of the central arteries of a zebrafish that has blood flow (**k**) versus an embryo in which blood flow never started through injection of the tnnt2a MO (**l**). Formation of the vasculature is largely normal if analyzed at early time points (images at 48 hpf) [47]. (**m–p**) Confocal images of *Tg(fli1:egfp)y1* wild-type in zebrafish silent heart mutants (*sih/sih*) demonstrating that the primary intersegmental vessels appear normally at 1.5 dpf (**m**), that secondary sprouts (arrows) and the parachordal line also develop normally (1.5–2.5 dpf; **n, o**), but that by ~3.5 dpf intersegmental vessels show enlargement of their dorsal segments (arrows) and collapsed vessels that are no longer lumenized (**p**) [48]. Images demonstrate the need to analyze vascular phenotypes as early in development as possible if the zebrafish do not have proper blood flow. Scale bars: 50 µm

embryos can show vascular defects in response to mis-patterned and delayed tissue development (Fig. 2a–f), emphasizing the take home message that phenotypes should be interpreted under properly controlled experimental conditions and phenotypic penetrance should be reported as part of studies.

1.3.3 Assessing Circulation and Cardiac Function

Defective blood flow (due for example to heart-specific defects) can affect the timing and extent of vessel growth, even in animals with apparently normal overall development. Assessing and confirming blood flow and cardiac function in the zebrafish is easily accomplished by direct imaging using a dissecting light microscope. For higher resolution, tracers injected into the blood stream such as quantum dots, fluorescent microspheres, lectins, and varying-molecular-weight fluorescent dextrans can be used to assess blood flow rates, blood flow directionality, vascular leak, vessel drainage sites, and solute uptake [49–52]. Although fish can survive a number of days in the absence of blood flow, animals lacking circulation do eventually become sick, edemic, stunted in growth, and die. If a mutant or morpholino experiment generates fish with no blood flow or decreased cardiac output, assessment of vascular phenotypes should take place as early as possible during development, during a time frame in which the animal is as healthy as possible. Although embryos and early larvae can develop reasonably normally for 1 or 2 days in the absence of circulation (Fig. 2k, l), overall vascular development becomes progressively disturbed over time (Fig. 2k–p). As a general rule, phenotypes in animals deficient in flow should be assessed prior to 6 days post fertilization (dpf), and ideally before 3 dpf, particularly when flow is entirely absent. As noted for animals with general developmental delay or morphological defects, caution should also be taken in interpreting vascular phenotypes observes in animals that lack circulatory flow. This is particularly important when assessing phenotypes such as lumen formation—as loss of a patient vasculature from lack of blood flow is not necessarily derived from the same molecular pathways as failure to ever open up luminal space during tube formation.

Quantitative assessment of blood flow and hemodynamics can be accomplished using particle velocimetry (PV), in which tracer particles injected into the blood stream are tracked through different vessels [53] and flow speed determined by calculating the displacement of the particles over time [54]. Due to depth and/or the decreased transparency at later stages, PV has mainly been used in embryos or early larvae, although recent advances in confocal microscopy, ultrasound, and tomography tools and methods have facilitated deeper imaging in more opaque tissues.

1.3.4 Assessing Vascular Patterning

Visualization and assessment of vascular patterning in the zebrafish is easily accomplished using widely available transgenic fluorescent reporter lines labeling vessels. Numerous transgenic zebrafish lines are available with cytoplasmic, nuclear, or membrane-localized fluorescent proteins expressed specifically in cardiovascular cells and tissues [55–57]. These include lines marking the heart endocardium and myocardium, blood and lymphatic vessel endothelium, hematopoietic derived cells, and perivascular cells like pericytes and fluorescent granular perithelial cells. Long-term time-lapse imaging methods have also been developed that permit continuous, real-time imaging of heart and vascular development in these transgenic lines [58]. Together, the availability of fluorescent transgenic reporter lines and methods for high-resolution, dynamic imaging of these lines has revolutionized the study of cardiovascular development.

Alternatively, or in addition, intravascular injection of fluorescent tracers such as quantum dots, fluorescent microspheres, or fluorescent dextrans, or nonfluorescent tracers such as India ink, Berlin blue, or Evans blue dyes can be used to visualize zebrafish blood vessels and their patterning [49, 51, 58–63]. Although the wide availability of vascular-specific transgenic lines has reduced the need for these traditional methods for examining the patterning of developing vessels, microangiography is still the method of choice for assessing vascular integrity, vessel lumenization, and flow (as noted above) in blood vessels, as well as permeability, drainage, and solute/fluid uptake in lymphatic vessels. The microangiographic techniques are fast, robust, and cost-effective methods for visualizing the vasculature and can be used to visualize vessels in nontransgenic animals or at more mature developmental stages.

The patterning of vessels in genetically or experimentally manipulated animals can be examined for alterations including reduced/absent vessel growth, excessive vessel growth or branching, altered patterning of vessels, and changes in flow patterns/vascular connections. Changes can occur either broadly throughout the animal or may be localized to particular vascular beds. Up to approximately 7 dpf the pattern of blood vessels observed should be compared to the published staged atlas of vascular anatomy [59]. Although the positioning of major vessels (e.g., dorsal aorta, cardinal vein, and intersegmental vessels in the trunk, or basilar artery in the head) is relatively invariant in normal animals at given stages, the precise location and paths taken by many smaller and/or later forming vessels (e.g., the central arteries of the hindbrain) is less stereotypic, and minor changes in the patterning of these vessels may not be significant. As noted above, it is important to assess vessel defects in genetically or experimentally manipulated animals in the context of the animal as a whole, including whether any developmental delay or gross morphological defects are present. The timing and extent of intersegmental vessel sprouting is frequently used to assess angiogenic phenotypes in developing zebrafish, but their patterning can be strongly affected by developmental delay, nonvascular morphological defects, and reduced or absent circulatory flow (Fig. 2).

1.4 Summary of Considerations in Assessing Vascular Phenotypes

The zebrafish remains a powerful tool for the study of vascular development. A variety of reverse-genetic approaches now exist in the fish for loss-of-function gene analysis, most notably morpholinos and CRISPR-generated genetic mutants. Although morpholinos have been questioned in recent years as a "first line" approach for assessment of loss-of-function phenotypes, they remain a valuable tool when validated by a corresponding genetic mutant. Furthermore, subsets of genetic mutants are susceptible to genetic compensation by upregulation of related genes, while morphants are not [29]. Detailed "community guidelines" for use of morpholinos were recently published and should be referred to when this tool is being utilized [1].

Additionally, researchers generating loss-of-function vascular phenotypes in the zebrafish should critically assess their data in light of some key questions:

- Where is the gene expressed?

- Do mutant or morphant phenotypes make sense when compared to the expression pattern of the gene? The more "vascular specific" an expression is, the more likely the associated vascular phenotypes are autonomous. Conversely, if a gene is ubiquitously expressed throughout the animal, the burden of proof is on the researcher to determine the vascular cell-autonomous function of the gene, using transplantation, tissue specific transgene "rescue" experiments, or other methods.

- Do mutants or morphants show significant developmental delay or gross morphological phenotypes, and do the vascular phenotypes reflect developmentally "younger" animals or problems in nonvascular tissues?

- Do the mutants or morphants have absent or strongly reduced circulatory flow?

- Are vessel phenotypes being assessed at a time point prior to lack of flow, which can cause significant general defects in development?

All these criteria should be taken into consideration to ensure reliable, valid conclusions when assessing vascular phenotypes in the zebrafish.

2 Materials

2.1 Microangiography

1. 0.02–0.04 μm fluoresceinated carboxylated latex beads, available from Invitrogen. The yellow-green (cat# F8787), red-orange (cat# F8794), or dark red (cat# F8783) beads are suitable for confocal imaging using the laser lines on standard Krypton-Argon laser confocal microscopes. Other colors may

be used for when multiphoton imaging is employed. Quantum dots (QD) are fluorescent semiconductors that are especially suited for multiphoton imaging with their high quantum yield, as well as the fact that the same excitation wavelength could be used for obtaining multiple different colors depending on the QD used. For microangiography, PEG-coated nontargeted QDs are available from Invitrogen. Qtracker 565, 655, 705, and 800 (cat# Q21031MP, Q21021MP, Q21061MP, and Q21071MP, respectively) are suitable for multiphoton confocal imaging. Fluorescent bead suspension as supplied is diluted 1:1 with 2% BSA (Sigma) in deionized distilled water, sonicated approximately 25 cycles of $1''$ each at maximum power on a Branson sonifier equipped with a microprobe, and subjected to centrifugation for 2 min at top speed in an Eppendorf microcentrifuge. The quantum dots are used as supplied.

2. 1 mm OD glass capillaries (World Precision Instruments, Cat# TW100-4 without filament or TW100F-4 with an internal filament) for preparing holding and microinjection pipettes.

3. 2 Coarse micromanipulators with magnetic holders and base plates.

4. $30\times$ Danieu's solution: 1740 mM NaCl, 21 mM KCl, 12 mM MgSO$_4$, 18 mM Ca(NO$_3$)$_2$, 150 mM HEPES, pH 7.6. Diluted to $1\times$ for use.

5. Holding and microinjection pipettes.

6. 6 cm culture dish (Falcon).

7. Micromanipulator and microinjection apparatus.

8. Dissecting microscope equipped with epifluorescence optics.

2.2 Long-Term Time-Lapse Imaging

1. 2% low-melting temperature agarose made up in $1\times$ Danieu's solution containing $1\times$ PTU (if nonalbino animals are used) and $1.25\times$ Tricaine (if nonparalyzed animals are used).

2. $1\times$ Danieu's solution with or without $1\times$ PTU and $1.25\times$ Tricaine (see above).

3. Fine forceps (Dumont #55).

4. Desired imaging chamber (*see* [63] for chamber suggestions).

3 Methods

3.1 Microangiography

Microangiography is useful for visualizing and assessing if developing vessels are actually carrying flow and/or have open lumens connected to a vasculature. This method allows for detailed study of both the pattern, function, and integrity (leakiness) of vessels during discrete stages of development. The method is straightforward to perform and does not require that the animal be of any particular genotype (although animals with impaired circulation may be difficult or impossible to infuse with fluorescent microspheres) [63].

1. The glass microinjection needles are prepared from 1 mm capilaries with internal filaments using a Kopf vertical pipette puller (approximate settings: heat = 12, solenoid = 4.5). Needles are broken open with a razor blade just behind their tip to give an opening of approximately 5–10 μm in width.

2. The holding pipettes (optional) are prepared from 1 mm capilaries WITHOUT filaments by carefully partially melting one end of the capillary with a Bunsen burner, such that the opening is narrowed to approximately 0.2 mm (*see* **Note 1**).

3. The apparatus for microinjection is made by attaching a glass microinjection needle (**step 1**) to a pipette holder (*see* **Note 2**). The pipette holder is attached to a controlled air pressure station such as World Precision Instruments Pneumatic Pico-pump (catalog # PV820).

4. The apparatus for holding embryos (optional) is made by attaching a glass holding pipette (**step 2**) to a pipette holder. The holding pipettes and their holders are attached via mineral-oil filled tubing to a manual microsyringe pump with 25 μL syringe (*see* **Note 3**).

5. Holding pipettes and microneedles and their associated holders and other equipment are arranged on either side of a stereo dissecting microscope.

1. Embryos are collected and incubated to the desired stage of development (*see* **Note 4**).

2. A few microliters of fluorescent microsphere suspension are used to backfill a glass microneedle for injection. The tip should be broken off to the desired diameter just before use.

3. Embryos are dechorionated and anesthetized with tricaine (MS-222) in embryo media.

4. 1–3-day-old embryos and larvae are held ventral side up for injection using a holding pipette applied to the side of the yolk ball, with suction applied via a microsyringe driver (*see* **Note 5**). 4–7-day-old larvae are held ventral side up for injection by embedding in 0.5% low melting temperature agarose.

5. For 1–3-day-old embryos a broken glass microneedle is inserted obliquely into the sinus venosus. For 4–7-day-old larvae a broken glass microneedle is inserted through the pericardium directly into the ventricle.

6. Following microneedle insertion, many (20+) small boluses of bead suspension are delivered over the course of up to a minute (*see* **Note 6**). The epifluorescence attachment on the dissecting microscope can be used to monitor the success of the injection.

7. Embryos or larvae are allowed to recover from injection briefly (approximately 1 min) in tricaine-free embryo media, then rapidly mounted in 5% methyl cellulose (Sigma) or 1.0–2.0% low-melt agarose (both made up in embryo media with tricaine). For short-term imaging (generally one stack of images) methylcellulose is applied to the bottom of a thick depression well slide (*see* **Note 7**). The rest of the well is carefully filled with 1× Danieu's solution containing 1× Tricaine, trying not to disturb the methylcellulose layer below. The injected zebrafish embryo is placed in the well, moved on top of the methylcellulose, and then gently pushed into the methylcellulose in the desired orientation to fully immobilize. For longer term or repeated imaging animals can be mounted in 1.0% low-melt agarose, using methods such as that described below.

8. Injected, mounted animals are imaged on a confocal or multi-photon microscope using the appropriate laser lines/wavelengths. Although the fluorescent beads are initially distributed uniformly throughout the vasculature of the embryo, within minutes they began to be phagocytosed by and concentrate into selected cells lining the vessels [64]; therefore, specimens must be imaged as rapidly as possible, generally within 15 min after injection.

3.2 Long-Term Time-Lapse Imaging

For time-lapse imaging of the vasculature using transgenic lines over the course of hours or even days, the animals must be carefully mounted in a way that maintains the region of the animal being imaged in a relatively fixed position, yet keeps the animal alive and developing normally throughout the course of the experiment. As the zebrafish embryos and larvae are continuously growing and undergoing morphogenetic movements, this must be accommodated in whatever scheme is used to hold them in place. Below, we outline a straightforward mounting method that is adaptable to imaging different areas of embryos or larvae and holds them in place over the course of hours. For time-lapse experiments that run up to a day or more imaging chambers with buffer circulation are employed.

3.2.1 Mounting Animals for Long-Term Time-Lapse Imaging

The following procedure is carried out using a dissecting microscope.

1. Dechorionate and anesthetize embryos with tricaine (MS-222) in embryo media. Select desired embryos for mounting. Only a single embryo is generally mounted per imaging chamber.

2. Fill the center of the imaging vessel with the 2% low-melting agarose/1× Danieu's solution to just below the rim of the chamber.

3. Using fine forceps blades, make a shallow, narrow trench in the center of the agarose dome where the fish will ultimately be oriented (*see* **Note 8**).

4. Two additional large cavities should also be carved out perpendicular to the trench on either side. These cavities will act as anchor points for the agarose layer that is overlaid on top of the embryo.

5. Place the embryo in the trench, and slowly overlay with molten agarose (*see* **Note 9**).

6. Cut excess agarose away by using the blades of a pair of fine forceps (*see* **Note 10**).

4   Notes

1. Slightly smaller for younger embryos, slightly larger for older larvae.

2. Adapter for holder and tubing to attach to picopump both from World Precision Instruments.

3. Pipette holder is from World Precision Instruments. Tubing and manual microsyringe pump are from Stoelting Instruments.

4. Use of albino mutant lines or PTU treatment improves visualization of many vascular beds at later stages.

5. Care should be taken to not allow the holding pipette to rupture the yolk ball.

6. Smaller numbers of overly large boluses can cause temporary or permanent cardiac arrest, and are therefore not recommended.

7. Methylcellulose is only useful for short-term mounting because the embryo gradually sinks as the methylcellulose loses viscosity from absorbing additional water over time.

8. This trench should be slightly wider than the dimensions of the embryo in its desired orientation for imaging [63]. It is critical that the trench be carved out carefully to make a space that holds the animal relatively motionless at rest. For imaging most vessels in the trunk of the fish the animal should lie on its side in the trench, for lateral view. If imaging is desired of vessels in the head, the embryo can be mounted laterally as described above, or vertically to look 'down' from the dorsal aspect onto the cranial vasculature. A larger cavity should be carved out posterior to the tail to accommodate additional increases in trunk/ tail length. Additional space should also be left around the head to accommodate shifting and growth, particularly in younger animals.

9. The agarose should be warm enough to freely flow in the buffer, but not too hot to kill the embryo. We typically use glass Pasteur pipette for this since it offers more precise control. Start from one cavity made in **step 4** next to the embryo trench. Apply the agarose at steady rate, filling the cavity. Then move over to the other cavity continuing to dispense agar over the embryo. Fill the opposing cavity up as well. This creates an agarose "bridge" over the embryo, holding it in place for imaging.

10. For imaging the trunk, we slice away triangles of agarose over the trunk and tail and over the rostral region, leaving a "bridge" of agarose over the yolk sac, posterior head, and anterior-most trunk sufficient to hold the embryo firmly in place but clears the desired sites of imaging from agarose, ensuring optical clarity of the trunk vessels. These cuts are necessary, since the embryo is growing in anterior-posterior axis, and straightening. Without removing these wedges of agarose continued growth and straightening of the embryo/larva is not accommodated.

References

1. Stainier DYR, Raz E, Lawson ND et al (2017) Guidelines for morpholino use in zebrafish. PLoS Genet 13(10):e1007000. https://doi.org/10.1371/journal.pgen.1007000

2. Kok FO, Shin M, Ni CW et al (2015) Reverse genetic screening reveals poor correlation between morpholino-induced and mutant phenotypes in zebrafish. Dev Cell 32(1):97–108. https://doi.org/10.1016/j.devcel.2014.11.018

3. Rossi A, Kontarakis Z, Gerri C et al (2015) Genetic compensation induced by deleterious mutations but not gene knockdowns. Nature 524(7564):230–233. https://doi.org/10.1038/nature14580

4. Lalonde S, Stone OA, Lessard S et al (2017) Frameshift indels introduced by genome editing can lead to in-frame exon skipping. PLoS One 12(6):e0178700. https://doi.org/10.1371/journal.pone.0178700

5. Prykhozhij SV, Steele SL, Razaghi B et al (2017) A rapid and effective method for screening, sequencing and reporter verification of engineered frameshift mutations in zebrafish. Dis Model Mech 10(6):811–822. https://doi.org/10.1242/dmm.026765

6. Varshney GK, Pei W, LaFave MC et al (2015) High-throughput gene targeting and phenotyping in zebrafish using CRISPR/Cas9. Genome Res 25(7):1030–1042. https://doi.org/10.1101/gr.186379.114

7. Zhang Y, Huang H, Zhang B et al (2016) TALEN- and CRISPR-enhanced DNA homologous recombination for gene editing in zebrafish. Methods Cell Biol 135:107–120. https://doi.org/10.1016/bs.mcb.2016.03.005

8. Meng X, Noyes MB, Zhu LJ et al (2008) Targeted gene inactivation in zebrafish using engineered zinc-finger nucleases. Nat Biotechnol 26(6):695–701. https://doi.org/10.1038/nbt1398

9. Villefranc JA, Amigo J, Lawson ND (2007) Gateway compatible vectors for analysis of gene function in the zebrafish. Dev Dyn 236 (11):3077–3087. https://doi.org/10.1002/dvdy.21354

10. Blackburn PR, Campbell JM, Clark KJ et al (2013) The CRISPR system--keeping zebrafish gene targeting fresh. Zebrafish 10 (1):116–118. https://doi.org/10.1089/zeb.2013.9999

11. Chang N, Sun C, Gao L et al (2013) Genome editing with RNA-guided Cas9 nuclease in zebrafish embryos. Cell Res 23(4):465–472. https://doi.org/10.1038/cr.2013.45

12. Hruscha A, Krawitz P, Rechenberg A et al (2013) Efficient CRISPR/Cas9 genome editing with low off-target effects in zebrafish.

Development 140(24):4982–4987. https://doi.org/10.1242/dev.099085

13. Hwang WY, Fu Y, Reyon D et al (2013) Heritable and precise zebrafish genome editing using a CRISPR-Cas system. PLoS One 8(7): e68708. https://doi.org/10.1371/journal.pone.0068708

14. Hwang WY, Fu Y, Reyon D et al (2013) Efficient genome editing in zebrafish using a CRISPR-Cas system. Nat Biotechnol 31 (3):227–229. https://doi.org/10.1038/nbt.2501

15. Jao LE, Wente SR, Chen W (2013) Efficient multiplex biallelic zebrafish genome editing using a CRISPR nuclease system. Proc Natl Acad Sci U S A 110(34):13904–13909. https://doi.org/10.1073/pnas.1308335110

16. Auer TO, Del Bene F (2014) CRISPR/Cas9 and TALEN-mediated knock-in approaches in zebrafish. Methods 69(2):142–150. https://doi.org/10.1016/j.ymeth.2014.03.027

17. Reade A, Motta-Mena LB, Gardner KH et al (2017) TAEL: a zebrafish-optimized optogenetic gene expression system with fine spatial and temporal control. Development 144 (2):345–355. https://doi.org/10.1242/dev.139238

18. Kimmel CB, Ballard WW, Kimmel SR et al (1995) Stages of embryonic development of the zebrafish. Dev Dyn 203(3):253–310. https://doi.org/10.1002/aja.1002030302

19. Butler D (2000) Wellcome Trust funds bid to unravel zebrafish genome. Nature 408 (6812):503. https://doi.org/10.1038/35046231

20. Westerfield M (1995) The zebrafish book. University of Oregon Press, Eugene, OR

21. Weinstein BM, Fishman MC (1996) Cardiovascular morphogenesis in zebrafish. Cardiovasc Res 31 Spec No:E17–E24

22. Weinstein BM, Schier AF, Abdelilah S et al (1996) Hematopoietic mutations in the zebrafish. Development 123:303–309

23. Mullins MC, Hammerschmidt M, Haffter P et al (1994) Large-scale mutagenesis in the zebrafish: in search of genes controlling development in a vertebrate. Curr Biol 4 (3):189–202

24. Driever W, Solnica-Krezel L, Schier AF et al (1996) A genetic screen for mutations affecting embryogenesis in zebrafish. Development 123:37–46

25. Stainier DY, Fouquet B, Chen JN et al (1996) Mutations affecting the formation and function of the cardiovascular system in the zebrafish embryo. Development 123:285–292

26. Burg L, Palmer N, Kikhi K et al (2018) Conditional mutagenesis by oligonucleotide-mediated integration of loxP sites in zebrafish. PLoS Genet 14(11):e1007754. https://doi.org/10.1371/journal.pgen.1007754

27. Gupta A, Hall VL, Kok FO et al (2013) Targeted chromosomal deletions and inversions in zebrafish. Genome Res 23(6):1008–1017. https://doi.org/10.1101/gr.154070.112

28. Liu P, Luk K, Shin M et al (2019) Enhanced Cas12a editing in mammalian cells and zebrafish. Nucleic Acids Res 47(8):4169–4180. https://doi.org/10.1093/nar/gkz184

29. El-Brolosy MA, Kontarakis Z, Rossi A et al (2019) Genetic compensation triggered by mutant mRNA degradation. Nature 568 (7751):193–197. https://doi.org/10.1038/s41586-019-1064-z

30. Lawson ND, Scheer N, Pham VN et al (2001) Notch signaling is required for arterial-venous differentiation during embryonic vascular development. Development 128 (19):3675–3683

31. Roman BL, Pham VN, Lawson ND et al (2002) Disruption of acvrl1 increases endothelial cell number in zebrafish cranial vessels. Development 129(12):3009–3019

32. Lawson ND, Mugford JW, Diamond BA et al (2003) phospholipase C gamma-1 is required downstream of vascular endothelial growth factor during arterial development. Genes Dev 17 (11):1346–1351. https://doi.org/10.1101/gad.1072203

33. Torres-Vazquez J, Gitler AD, Fraser SD et al (2004) Semaphorin-plexin signaling guides patterning of the developing vasculature. Dev Cell 7(1):117–123. https://doi.org/10.1016/j.devcel.2004.06.008

34. Shaw KM, Castranova DA, Pham VN et al (2006) fused-somites-like mutants exhibit defects in trunk vessel patterning. Dev Dyn 235(7):1753–1760. https://doi.org/10.1002/dvdy.20814

35. Pham VN, Lawson ND, Mugford JW et al (2007) Combinatorial function of ETS transcription factors in the developing vasculature. Dev Biol 303(2):772–783. https://doi.org/10.1016/j.ydbio.2006.10.030

36. Gore AV, Swift MR, Cha YR et al (2011) Rspo1/Wnt signaling promotes angiogenesis via Vegfc/Vegfr3. Development 138 (22):4875–4886. https://doi.org/10.1242/dev.068460

37. Pan W, Pham VN, Stratman AN et al (2012) CDP-diacylglycerol synthetase-controlled phosphoinositide availability limits VEGFA signaling and vascular morphogenesis. Blood

120(2):489–498. https://doi.org/10.1182/blood-2012-02-408328

38. Moens CB, Donn TM, Wolf-Saxon ER et al (2008) Reverse genetics in zebrafish by TILLING. Brief Funct Genom Proteom 7 (6):454–459. https://doi.org/10.1093/bfgp/eln046

39. Varshney GK, Carrington B, Pei W et al (2016) A high-throughput functional genomics workflow based on CRISPR/Cas9-mediated targeted mutagenesis in zebrafish. Nat Protoc 11 (12):2357–2375. https://doi.org/10.1038/nprot.2016.141

40. Labun K, Guo X, Chavez A et al (2019) Accurate analysis of genuine CRISPR editing events with ampliCan. Genome Res 29(5):843–847. https://doi.org/10.1101/gr.244293.118

41. Labun K, Montague TG, Krause M et al (2019) CHOPCHOP v3: expanding the CRISPR web toolbox beyond genome editing. Nucleic Acids Res 47(W1):W171–W174. https://doi.org/10.1093/nar/gkz365

42. Montague TG, Cruz JM, Gagnon JA et al (2014) CHOPCHOP: a CRISPR/Cas9 and TALEN web tool for genome editing. Nucleic Acids Res 42(Web Server issue):W401–W407. https://doi.org/10.1093/nar/gku410

43. Hoshijima K, Jurynec MJ, Klatt Shaw D et al (2019) Highly efficient CRISPR-Cas9-based methods for generating deletion mutations and F0 embryos that lack gene function in zebrafish. Dev Cell 51(5):645–657.e644. https://doi.org/10.1016/j.devcel.2019.10.004

44. Won M, Ro H, Dawid IB (2015) Lnx2 ubiquitin ligase is essential for exocrine cell differentiation in the early zebrafish pancreas. Proc Natl Acad Sci U S A 112(40):12426–12431. https://doi.org/10.1073/pnas.1517033112

45. Cha YR, Weinstein BM (2012) Use of PCR template-derived probes prevents off-target whole mount in situ hybridization in transgenic zebrafish. Zebrafish 9(2):85–89. https://doi.org/10.1089/zeb.2011.0731

46. Yu JA, Castranova D, Pham VN et al (2015) Single-cell analysis of endothelial morphogenesis in vivo. Development 142 (17):2951–2961. https://doi.org/10.1242/dev.123174

47. Fujita M, Cha YR, Pham VN et al (2011) Assembly and patterning of the vascular network of the vertebrate hindbrain. Development 138(9):1705–1715. https://doi.org/10.1242/dev.058776

48. Isogai S, Lawson ND, Torrealday S et al (2003) Angiogenic network formation in the developing vertebrate trunk. Development 130(21):5281–5290. https://doi.org/10.1242/dev.00733

49. Yaniv K, Isogai S, Castranova D et al (2006) Live imaging of lymphatic development in the zebrafish. Nat Med 12(6):711–716. https://doi.org/10.1038/nm1427. nm1427 [pii]

50. Isogai S, Hitomi J, Yaniv K et al (2009) Zebrafish as a new animal model to study lymphangiogenesis. Anat Sci Int 84(3):102–111. https://doi.org/10.1007/s12565-009-0024-3

51. Venero Galanternik M, Castranova D, Gore AV et al (2017) A novel perivascular cell population in the zebrafish brain. elife 6. https://doi.org/10.7554/eLife.24369

52. Jung HM, Castranova D, Swift MR et al (2017) Development of the larval lymphatic system in the zebrafish. Development 144:2070. https://doi.org/10.1242/dev.145755

53. Poelma C, Kloosterman A, Hierck BP et al (2012) Accurate blood flow measurements: are artificial tracers necessary? PLoS One 7(9): e45247. https://doi.org/10.1371/journal.pone.0045247

54. Vennemann P, Kiger KT, Lindken R et al (2006) In vivo micro particle image velocimetry measurements of blood-plasma in the embryonic avian heart. J Biomech 39 (7):1191–1200. https://doi.org/10.1016/j.jbiomech.2005.03.015

55. Kwan KM, Fujimoto E, Grabher C et al (2007) The Tol2kit: a multisite gateway-based construction kit for Tol2 transposon transgenesis constructs. Dev Dyn 236(11):3088–3099. https://doi.org/10.1002/dvdy.21343

56. Kawakami K, Abe G, Asada T et al (2010) zTrap: zebrafish gene trap and enhancer trap database. BMC Dev Biol 10:105. https://doi.org/10.1186/1471-213X-10-105

57. Zhang Y, Werling U, Edelmann W (2012) SLiCE: a novel bacterial cell extract-based DNA cloning method. Nucleic Acids Res 40 (8):e55. https://doi.org/10.1093/nar/gkr1288

58. Kamei M, Isogai S, Pan W et al (2010) Imaging blood vessels in the zebrafish. Methods Cell Biol 100:27–54. https://doi.org/10.1016/B978-0-12-384892-5.00002-5

59. Isogai S, Horiguchi M, Weinstein BM (2001) The vascular anatomy of the developing zebrafish: an atlas of embryonic and early larval development. Dev Biol 230(2):278–301. https://doi.org/10.1006/dbio.2000.9995

60. Lawson ND, Weinstein BM (2002) In vivo imaging of embryonic vascular development using transgenic zebrafish. Dev Biol 248 (2):307–318. S0012160602907116 [pii]

61. Stratman AN, Pezoa SA, Farrelly OM et al (2017) Interactions between mural cells and endothelial cells stabilize the developing zebrafish dorsal aorta. Development 144 (1):115–127. https://doi.org/10.1242/dev.143131

62. Choi WY, Gemberling M, Wang J et al (2013) In vivo monitoring of cardiomyocyte proliferation to identify chemical modifiers of heart regeneration. Development 140(3):660–666. https://doi.org/10.1242/dev.088526

63. Jung HM, Isogai S, Kamei M et al (2016) Imaging blood vessels and lymphatic vessels in the zebrafish. Methods Cell Biol 133:69–103. https://doi.org/10.1016/bs.mcb.2016.03.023

64. Westerfield M (2000) The zebrafish book. A guide for the laboratory use of zebrafish (Danio rerio), 4th edn. Univ. of Oregon Press, Eugene, OR

Chapter 16

By the Skin of Your Teeth: A Subcutaneous Mouse Model to Study Pulp Regeneration

Annelies Bronckaers, Petra Hilkens, Esther Wolfs, and Ivo Lambrichts

Abstract

Exiting developments in tissue engineering and new insights in stem cell biology have led to new possible strategies for the regeneration of damaged tissues in the oral cavity. The regeneration of the pulp–dentin complex regeneration in particular, has drawn the attention of many researchers because of the high clinical needs. While it is still important to perform in vitro research using a wide variety of cells, scaffolds and growth factors, it is also critical to have a reliable animal model for preclinical trials. In this chapter, we describe a mouse model in which a scaffold resembling a tooth containing dental pulp cells is implanted subcutaneously. We also describe which histological stainings could be used to examine blood vessel formation and the regeneration of the pulp–dentin complex.

Key words Pulp regeneration, Mouse model, Dental stem cells, Histology, Masson's trichrome, Tissue engineering

1 Introduction

Teeth have essential roles in numerous daily tasks, particularly in food intake and speech. Tooth loss can be caused by traumatic injuries but can also be the result of periodontal pathologies and caries which are triggered by deficient oral hygiene, poor nutrition, and smoking [1]. According to the World Health Organization (WHO), periodontitis is the sixth most prevalent health condition worldwide [2]. Unfortunately, human adult teeth have a limited regenerative potential. Current therapies for the replacement of missing teeth mostly use partial or complete dentures and titanium implants capped with a ceramic crown. Despite the fact that these prostheses serve a purpose, they are prone to mechanical and biological failure. Especially the restoration of tooth pulp is a much-investigated goal since the current practice of replacing infected pulp with inorganic materials (such as cement) results in a necrotic tooth [3]. There is thus an urgent need for new treatments that can aid or induce pulp regeneration.

Domenico Ribatti (ed.), *Vascular Morphogenesis: Methods and Protocols*, Methods in Molecular Biology, vol. 2206, https://doi.org/10.1007/978-1-0716-0916-3_16, © Springer Science+Business Media, LLC, part of Springer Nature 2021

The discovery of (dental) stem cells and novel advances in the field of tissue engineering (such as bioprinting and the development of new extracellular matrix materials), has led to an exciting boost in research on the regeneration of dental tissues [4, 5]. However, few preclinical experimental models are available for mechanistic and translational studies involving pulp regeneration. A very common model is a pulpotomy or pulpectomy rat model, in which the pulp is (partially) removed and next, the cells, growth factors or matrix materials (i.e., the treatment) are placed into the pulp cavity after which the tooth is closed. However, this is a very difficult model as it requires a lot of training and surgical precision [6, 7]. An excellent alternative are pig pulpectomy models, but these are not available for each research institute [8, 9]. Hence, we describe here an easier and more accessible mouse model for pulp regeneration: a "tooth slice/scaffold model of dental pulp tissue engineering," in which scaffolds or tooth slices are loaded with dental stem cells in a hydrogel and subcutaneously implanted into nude mice [9–11]. Of course, depending on the research question, growth factors or different types of stem cells can be added into the hydrogel, but is it is also possible to use other types of hydrogels and scaffolds. We and others have demonstrated that it is possible to use a synthetic, self-assembling peptide hydrogel gel to support dental pulp stem cells [11, 12]. PuraMatrix was chosen as a hydrogel as this is a synthetic hydrogel than can solidify at room temperature (after contact with ions which are present in saline solutions but also blood) and which is also available for clinical use. After subcutaneous transplantation in nude mice for several weeks, tissues can be harvested and processed for histological stainings [11]. The presence of blood vessels can be demonstrated by using traditional hematoxylin–eosin staining. Masson's trichrome staining, which is a connective tissue stain, is performed to identify the newly formed collagen which is synthesized by the transplanted stem cells adjacent to the scaffold.

2   Materials

1. Dental pulp (stem cells).

2. Forceps.

3. Ethanol.

4. Forceps or Splitter.

5. Cell culture medium: Minimal Essential Medium Eagle, alpha modification (αMEM) supplemented with 2 mM L-glutamine, 100 U/ mL penicillin, 100 µg/ mL streptomycin (Sigma-Aldrich), and 10% fetal bovine serum [13].

6. Trypsin–EDTA solution (preferentially 0.05% trypsin/EDTA).

7. Phosphate buffered saline solution, pH 7.4 (PBS).

8. 6-well dishes.

9. 15 and 50 mL tubes.

10. 25 cm^2 flasks.

11. CO_2 incubator (humidified, 5% CO_2 atmosphere, 37 °C).

12. Phase-contrast microscope.

13. Centrifuge.

14. Biosafety cabinet in L2 facility.

15. Corning® PuraMatrix™ Peptide Hydrogel, 5 mL, product number 354250, this consists of standard amino acids (1% w/v) and 99% water. Under physiological conditions, the peptide component self-assembles into a 3D hydrogel with a nanometer-scale fibrous structure. The hydrogel is readily formed in a Falcon culture dish, plate, or cell culture insert.

16. Scaffolds: These can either be scaffolds from hydroxyapatite or 1-mm-thick human tooth slices cross-sections from the cervical region of recently extracted noncarious human third molars [9, 10]. All scaffolds and tooth slices need to be sterile.

17. Scalpel.

18. Petri dishes.

19. Eppendorf tubes.

20. Mice: immunodeficient mice, preferentially nude mice (SCID are also possible, but they need to be shaved; alternatively, hairless SCID mice can be used such as hairless outbred (SHO™) mice (SHO-Prkdc^{scid}Hr^{hr}, Charles River)). These mice need to be housed in SPF facility.

21. Heating pad.

22. Fine iris scissors (Straight, 10.5 cm).

23. Standard scissors (straight blunt/blunt 12 cm).

24. Forceps.

25. Wound clips and wound clip applier.

26. NaCl or eye gel.

27. Inhaled isoflurane (2.5%).

28. Iodide solation (e.g., Iso-betadine 10%).

29. 4% paraformaldehyde (PFA).

30. Ethanol (70–80–90–95–100%).

31. Xylene.

32. Paraffin (e.g., Paraplast Plus Leica Biosystems).

33. Oven or heating baths.

34. Paraffin molds.

35. Surgical saw.

36. Decalcifying solution (Thermo Scientific™, Richard-Allan Scientific, Waltham, MA, USA).

37. Microtome.

38. A stereomicroscope equipped with a digital camera.

39. Histological baths.

40. Magnetic stirrer with heating pad.

41. Mounting medium (DPX).

42. Coverslips.

43. Mayer's hematoxylin.

44. Fuchsine.

45. Ponceau de xylidine.

46. Aniline Blue.

47. Glacial acetic acid.

48. Phosphomolybdic acid solution (1%).

49. Acetic acid solution (1%).

50. Eosin Y solution (2%).

3 Methods

3.1 Isolation and Culture of Dental Pulp Stem Cells (DPSC)

1. Immediately after extraction, sterilize the tooth with 100% ethanol.

2. Next, remove periodontal ligaments and root apical papilla from the root surface with a sterile forceps (*see* **Note 1**).

3. If the apex is not yet fully grown, collect the dental pulp out of the tooth with a forceps. If the pulp is not accessible, crack the tooth open with a tooth splitter.

4. In either case, emerge the pulp in 10 mL α-MEM (in a 50 mL tube) until arrival at the cell culture facility (within 3 h).

5. From this step onward, sterile techniques are necessary and working under a biosafety cabinet is required. Rinse the pulp 3 times with alpha-MEM and place them into an Eppendorf tubes. Cut the pulp into small pieces of maximum 1 mm^3 using sterile scissors.

6. Place all pieces in 2 mL of complete cell culture medium in a well of a 6-well plate and incubate them in a CO_2 incubator.

7. Leave them untouched for 48–72 h.

8. 48–72 h later, pulp tissues are attached to the plastic and cells start to grow out.

9. Replace medium twice a week.

10. On reaching 80–90% confluence, cells need to be transferred to another container. Remove the medium and wash the cells with PBS. Add 1 mL of trypsin/EDTA solution and at 37 °C for 3 min. When cells are detached, enzymatic action is inactivated by addition of compete cell culture medium which contains 10% fetal calf serum.

11. Transfer the cells into a 15 mL tube and centrifuge at $250 \times g$ (~150 rcf) for 5 min.

12. Remove the supernatant and resuspend the cells in complete cell culture medium.

13. Subcultivate cells to a 25 cm^2 flask at a ratio of 100,000 cells/ 25 cm^2.

14. Dental pulp stem cells are used until passage 7.

15. To characterize the DPSC, test the expression of the following (stem) cell markers by means of flow cytometry as described previously: negative for CD31, CD34 and CD45, while CD29, CD73, CD90, and CD105 have to be positive [14, 15].

3.2 Preparation of the Scaffold

1. Harvest dental stem cells as described above.

2. Wash the cells twice with PBS.

3. Count the number of cells and make a suspension in PBS in the preferred volume (e.g., 25,000 cells/50 μL; the total volume depends on the size of the scaffold used and needs to be optimized).

4. Dilute the PuraMatrix 1% solution to a 0.4% dilution in PBS. Mix equal **amounts** of the PuraMatrix with the cell solution to obtain a 0.2% PuraMatrix solution.

5. Work fast and while the cell–hydrogel solution is still liquid add the appropriate volume to the scaffold. The PuraMatrix–cell solution will solidify in the petri dish.

3.3 Implantation of the Scaffolds

1. Anesthetize the mouse with 2.5% isoflurane for induction and maintenance of anesthesia.

2. Check the depth of anesthesia by testing the paw pinch reflex; if the reflex is still present, wait a few more minutes before testing again.

3. Place the mouse onto a warm heating pad.

4. Add NaCl or eye gel to the eyes to prevent them from drying.

5. Wipe the dorsal skin with an iodide solation (e.g., Iso-betadine 10%) followed by 70% ethanol in order to disinfect it.

6. Make a 1.5–2 cm incision into the skin parallel to the longitudinal body axis of the mouse with a standard blunt scissor.

7. Place the closed blunt scissors within the incision and open them to loosen the skin from the tissue underneath—repeat this 3 times.

8. Place the scaffold with the opening in caudal direction within the incision with forceps.

9. Close the skin with wound clips.

10. Mark the mouse with ear cuts to be able to identify each individual mouse (*see* **Note 2**).

11. Place the mouse in a cage with extra bedding material for recovery.

12. Check the mouse daily for any discomfort, infection of the wound and whether the wound clips come off.

3.4 Tissue Isolation and Embedding into Paraffin

1. Eight or 12 weeks after transplantation, the scaffolds can be resected.

2. Euthanize the mice (*see* **Note 3**).

3. Wipe the dorsal skin with 70% ethanol.

4. Carefully lift the skin around the scaffold and make incision in the skin with blunt scissors.

5. Remove the scaffold without removing the hydrogel or the tissue that has grown inward. Figure 1b shows a macroscopic picture of a 3D scaffold shortly after surgical removal.

6. Immediately place the scaffold in 4% paraformaldehyde for 48 h at room temperature to fix the tissue.

7. When using large scaffolds, it might be useful to remove a piece of scaffold with a surgical saw, to allow complete impregnation of the formed tissue in paraffin.

8. Dehydrate the cylinders in graded alcohol (70% ethanol, 90% ethanol, two baths of 100% for each at least 1 h), clear them in xylene (3 times 1.5 h), and submerge them in paraffin wax (58–60 °C, two changes for 2 h).

9. Embed tissue into paraffin blocks, in the right orientation. When using scaffolds with only one opening, the opening must be oriented toward the surface of the block.

10. Paraffin-embedded constructs can then be decalcified with decalcifying solution for 24 h at room temperature. The time needed depends on the type of scaffold and need to be checked regularly (e.g., by testing the hardness with a needle). It is important that the scaffold is very close toward the surface. If this is not the case, the paraffin block should be trimmed until the scaffold reaches the surface. Decalcification can also be performed before embedment in paraffin but this can result in the newly formed tissue to collapse.

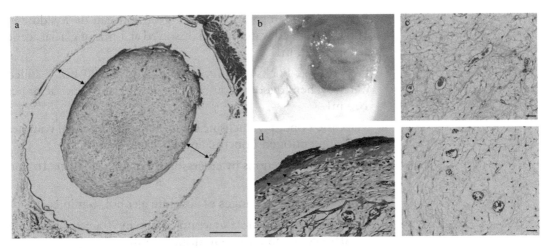

Fig. 1 Subcutaneous mouse model of pulp regeneration. 3D Hydroxyapatite constructs were loaded with dental pulp stem cells in PuraMatrix and implanted subcutaneously in scid mice. 12 weeks after transplantation, the scaffolds were removed. (**a**) Overview of TCM stained scaffold. Black arrows indicate the borders of the former scaffold. Scale bar = 500 μm; (**b**) Macroscopic view of the scaffold, shortly after resection. (**c**) Details of blood vessels in a TCM-stained scaffold, Scale bar = 50 μm. (**d**) The border of the tissue that came in close contact with the scaffolds, showed strongly organized, concentric layers of collagen (in bright blue) and, in certain cases, mineralized tissue (in red). Black arrows indicate the collagen layer. Scale bar = 50 μm; (**e**) Details of blood vessels in a H&E-stained scaffold, Scale bar = 50 μm

11. Next, allow the paraffin to melt again (at 60 °C) to allow better impregnation of the decalcified constructs.

12. 24 h later, bring the tissue into a mold with melted paraffin, and allow it to solidify when having the tissue in the right orientation.

13. Make serial sections of 7 μm and place the paraffin sections on an object glass.

14. These sections can be used for histological and immunological stainings; however, due to the calcification steps it can be possible that not all antigens remain preserved and that depending on the target, immunohistochemistry is not possible. General morphology can be analyzed by performing a hematoxylin–eosin staining.

3.5 Masson's Trichrome Staining

Masson's trichrome staining is a connective tissue stain, which is performed to identify the newly formed collagen which is synthesized by the transplanted stem cells adjacent to the scaffold (Fig. 1a, c, d).

1. Prepare 2% Aniline Blue solution: add 2 g of aniline blue to 100 mL of distilled water and allow the water to boil for 2 min. Next, after the solution is again at room temperature, add 2 mL of glacial acetic acid per 100 mL. Filter before use.

2. Prepare Ponceau solution: add 1 g of Ponceau de Xylidine in 100 mL distilled water. When dissolved at 1 mL of glacial acetic acid per 100 mL.

3. Prepare acid Fuchsine solution: add 1 g in 100 mL distilled water. When dissolved at 1 mL of glacial acetic acid per 100 mL.

4. Make the Ponceau/Fuchsine solution by mixing equal volumes of aforementioned Ponceau and Fuchsine solution.

5. Deparaffinize sections by emerging them in 2 times for 10 min in xylene.

6. Rehydrate in two changes of absolute alcohol, 5 min each.

7. Incubate for 2 min in 95% alcohol, 2 min in 90% ethanol, 2 min in 70% alcohol and 5 min in distilled water.

8. Incubate the slices for 10 min in Mayer's hematoxylin.

9. Incubate for 10 min in tap water.

10. Incubate for 5 min in Ponceau/Fuchsine solution and wash for 2 min in distilled water.

11. Differentiate in phosphomolybdic acid solution (1%) for 5 min.

12. Rinse in distilled water.

13. Incubate in Aniline Blue solution for 5–10 min and next for 2 min in distilled water.

14. Differentiate for the second time in phosphomolybdic acid solution (1%) for 5 min and rinse shortly in distilled water.

15. Incubate slide in acetic acid solution (1%) for 2 min and rinse with distilled water.

16. Dehydrate the slides by incubating them subsequently in 70% ethanol, 90% ethanol, 95% ethanol, and two baths of absolute alcohol, 5 min each.

17. Clear in two changes of xylene, 5 min each.

18. Mount with xylene based mounting medium (such as DPX).

3.6 Hematoxylin–Eosin

Alternatively, slides can also be stained with a general hematoxylin–eosin stain (Fig. 1e).

1. Deparaffinize sections by emerging them in 2 times for 10 min in xylene.

2. Rehydrate in two changes of absolute alcohol, 5 min each.

3. Incubate for 2 min in 95% alcohol, 2 min in 90% ethanol, 2 min in 70% alcohol, and 5 min in distilled water.

4. Incubate the slices for 8 min in Mayer's hematoxylin.

5. Incubate for 10 min in tap water to allow the stain to develop.

6. Incubate for 30–60 s in eosin solution.

7. Rapidly dehydrate the slides by incubating them subsequently in 70% ethanol, 90% ethanol, 95% ethanol, and two baths of absolute alcohol, 30 s each.

8. Clear in two changes of xylene, 5 min each.

9. Mount with xylene based mounting medium (such as DPX).

4   Notes

1. Remove all soft tissues that are attached to the outside of the tooth.

2. This applies when mice are housed together.

3. Not with cervical dislocation, preferentially with an overdose of pentobarbital.

References

1. Tyrovolas S, Koyanagi A, Panagiotakos DB et al (2016) Population prevalence of edentulism and its association with depression and self-rated health. Sci Rep 6:37083. https://doi.org/10.1038/srep37083

2. Kassebaum NJ, Bernabe E, Dahiya M et al (2014) Global burden of severe periodontitis in 1990-2010: a systematic review and meta-regression. J Dent Res 93:1045–1053. https://doi.org/10.1177/0022034514552491

3. Volponi AA, Pang Y, Sharpe PT (2010) Stem cell-based biological tooth repair and regeneration. Trends Cell Biol 20-206 (12-6):715–722. https://doi.org/10.1016/j.tcb.2010.09.012

4. Kim S, Shin SJ, Song Y et al (2015) In vivo experiments with dental pulp stem cells for pulp-dentin complex regeneration. Mediat Inflamm 2015:409347. https://doi.org/10.1155/2015/409347

5. Angelova Volponi A, Zaugg LK, Neves V et al (2018) Tooth repair and regeneration. Curr Oral Health Rep 5:295–303. https://doi.org/10.1007/s40496-018-0196-9

6. Lopes CS, Junqueira MA, Cosme-Silva L et al (2019) Initial inflammatory response after the pulpotomy of rat molars with MTA or ferric sulfate. J Appl Oral Sci 27:e20180550. https://doi.org/10.1590/1678-7757-2018-0550

7. Okamoto M, Takahashi Y, Komichi S et al (2018) Dentinogenic effects of extracted dentin matrix components digested with matrix metalloproteinases. Sci Rep 8:10690. https://doi.org/10.1038/s41598-018-29112-3

8. Zhu X, Liu J, Yu Z et al (2018) A miniature swine model for stem cell-based de novo regeneration of dental pulp and dentin-like tissue. Tissue Eng Part C Methods 24:108–120. https://doi.org/10.1089/ten.tec.2017.0342

9. Sakai VT, Cordeiro MM, Dong Z et al (2011) Tooth slice/scaffold model of dental pulp tissue engineering. Adv Dent Res 23:325–332. https://doi.org/10.1177/0022034511405325

10. Cordeiro MM, Dong Z, Kaneko T et al (2008) Dental pulp tissue engineering with stem cells from exfoliated deciduous teeth. J Endod 34:962–969. https://doi.org/10.1016/j.joen.2008.04.009

11. Hilkens P, Bronckaers A, Ratajczak J et al (2017) The angiogenic potential of DPSCs and SCAPs in an In Vivo model of dental pulp regeneration. Stem Cells Int 2017:2582080. https://doi.org/10.1155/2017/2582080

12. Dissanayaka WL, Hargreaves KM, Jin L et al (2015) The interplay of dental pulp stem cells and endothelial cells in an injectable peptide hydrogel on angiogenesis and pulp regeneration in vivo. Tissue Eng Part A 21:550–563. https://doi.org/10.1089/ten.TEA.2014.0154

13. Bronckaers A, Hilkens P, Fanton Y et al (2013) Angiogenic properties of human dental pulp stem cells. PLoS One 8:e71104. https://doi.org/10.1371/journal.pone.0071104

14. Gronthos S, Brahim J, Li W et al (2002) Stem cell properties of human dental pulp stem cells. J Dent Res 81:531–535. https://doi.org/10.1177/154405910208100806

15. Hilkens P, Gervois P, Fanton Y et al (2013) Effect of isolation methodology on stem cell properties and multilineage differentiation potential of human dental pulp stem cells. Cell Tissue Res 353:65–78. https://doi.org/10.1007/s00441-013-1630-x

Domenico Ribatti (ed.), *Vascular Morphogenesis: Methods and Protocols*, Methods in Molecular Biology, vol. 2206,
https://doi.org/10.1007/978-1-0716-0916-3, © Springer Science+Business Media, LLC, part of Springer Nature 2021